STUDENT EDITION

Grade 5

VOLUME 2

Mission 5: Volume, Area, and Shapes

Mission 6: The Coordinate Plane

NAME ______________________________

Printed in the U.S.A.

ISBN: 979-8-88868-889-2

Table of Contents

Mission 5

Lesson 1 3
Lesson 2 9
Lesson 3 15
Lesson 4 19
Lesson 5 23
Lesson 6 27
Lesson 7 29
Lesson 8 33
Lesson 9 37
Lesson 10 43
Lesson 11 47
Lesson 12 51
Lesson 13 57
Lesson 14 59
Lesson 15 63
Lesson 16 69
Lesson 17 73
Lesson 18 77
Lesson 19 81
Lesson 20 85
Lesson 21 89

Mission 6

Lesson 1 95
Lesson 2 97
Lesson 3 101
Lesson 4 105
Lesson 5 111
Lesson 6 115
Lesson 7 121
Lesson 8 127
Lesson 9 131
Lesson 10 135
Lesson 11 139
Lesson 12 143
Lesson 13 147
Lesson 14 149
Lesson 15 153
Lesson 16 155
Lesson 17 159
Lesson 18 161
Lesson 19 165
Lesson 20 169
Lesson 26 175
Lesson 27 179
Lesson 28 181

Lesson 29 185

Lesson 30 187

Lesson 31 189

Lesson 32 191

Lesson 33 197

Lesson 34 199

Grade 5

Mission 5

Volume, Area, and Shapes

CENTIMETER GRID PAPER (CONCEPT EXPLORATION TEMPLATE 1)

ISOMETRIC DOT PAPER (CONCEPT EXPLORATION TEMPLATE 2)

PROBLEM SET

1. Shade the following figures on centimeter grid paper. Cut and fold each to make 3 open boxes, taping them so they hold their shapes. Pack each box with cubes. Write how many cubes fill each box.

a.

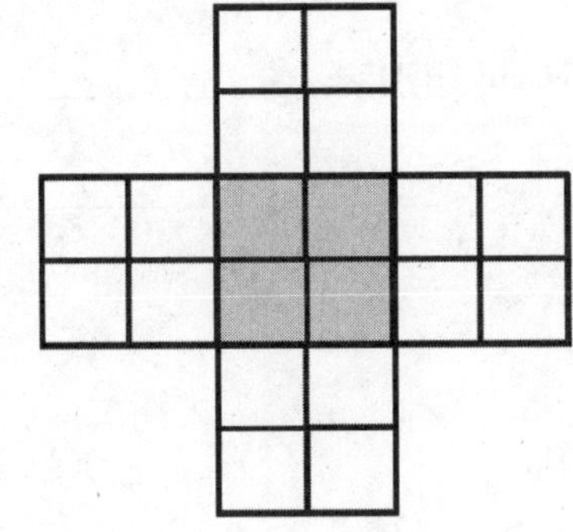

Number of cubes: ________________

b.

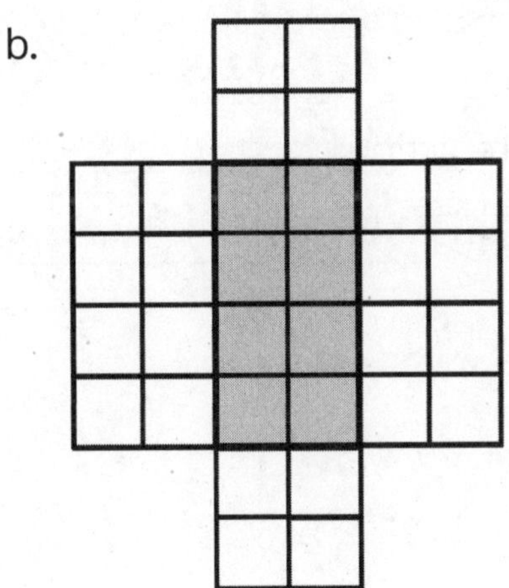

Number of cubes: ________________

c.

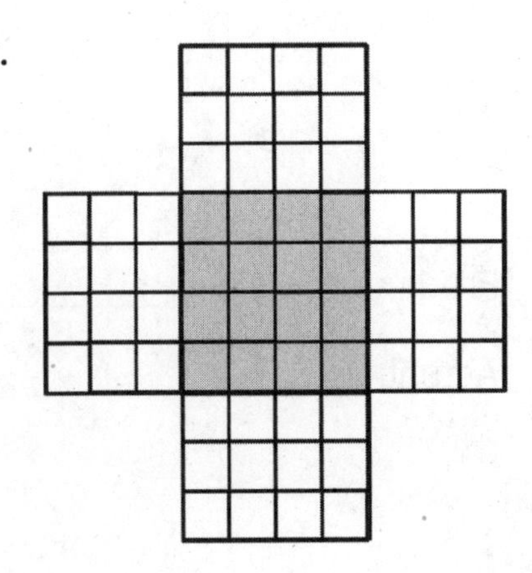

Number of cubes: ________________

2. Predict how many centimeter cubes will fit in each box, and briefly explain your predictions. Use cubes to find the actual volume. (The figures are not drawn to scale.)

a.

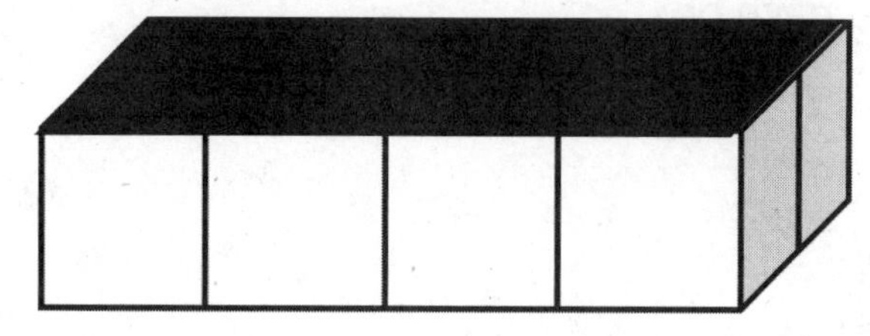

Prediction: ______________

Actual: ______________

b.

Prediction: ______________

Actual: ______________

c.

Prediction: ______________

Actual: ______________

3. Cut out the net in the template, and fold it into a cube. Predict the number of 1-centimeter cubes that would be required to fill it.

   a. Prediction: ____________________

   b. Explain your thought process as you made your prediction.

   c. How many 1-centimeter cubes are used to fill the figure? Was your prediction accurate?

RECTANGULAR PRISM RECORDING SHEET (CONCEPT EXPLORATION TEMPLATE)

RECTANGULAR PRISM RECORDING SHEET (CONCEPT EXPLORATION TEMPLATE)

Lesson 5

Name: ______________________ Date: ____________

GRADE 5 / MISSION 5 / LESSON 5

Exit Ticket

1. Find the volume of the prism.

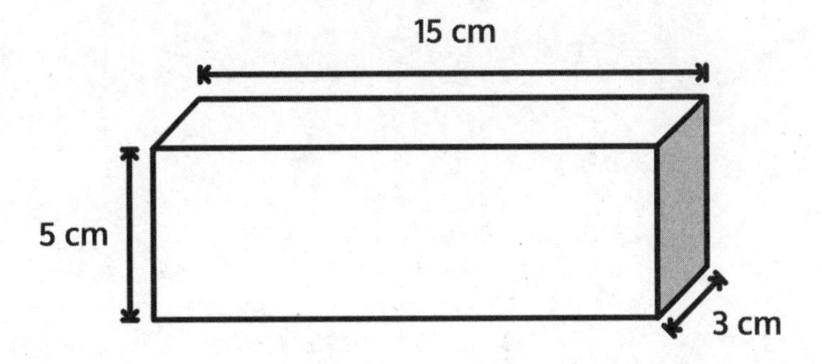

2. Shade the beaker to show how much liquid would fill the box.

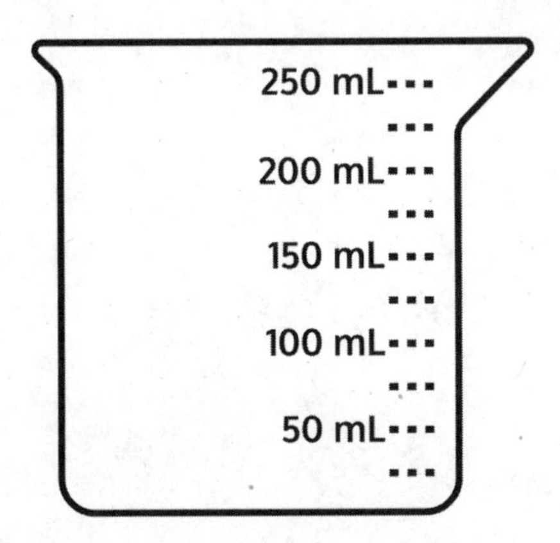

PROBLEM SET

1. Determine the volume of two boxes on the table using cubes, and then confirm by measuring and multiplying.

Box Number	Number of Cubes Packed	Measurements			Volume
		Length	Width	Height	

2. Using the same boxes from Problem 1, record the amount of liquid that your box can hold.

Box Number	Liquid the Box Can Hold
	mL
	mL

3. Shade to show the water in the beaker.

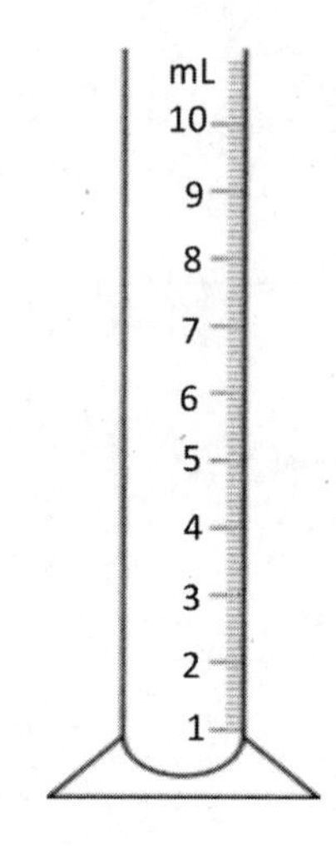

At first:

__________ mL

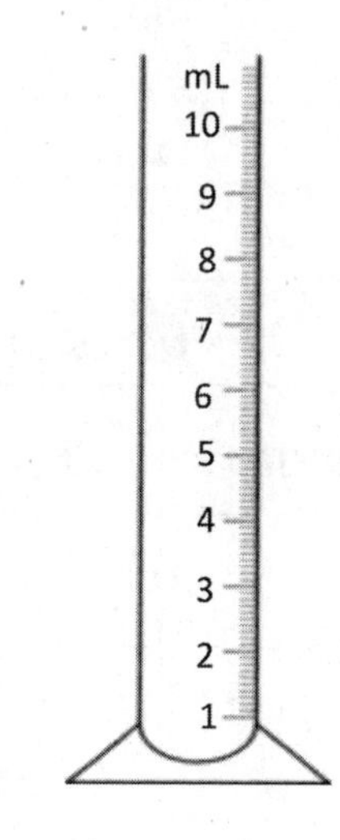

After 1 mL water added:

__________ mL

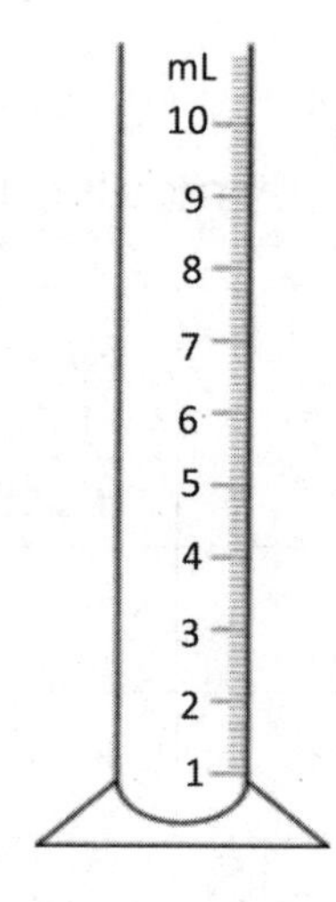

After 1 cm cube added:

__________ mL

Lesson 6

Word Problem

A storage company advertises three different choices for all your storage needs: "The Cube," a true cube with a volume of 64 m^3; "The Double" (double the volume of "The Cube"); and "The Half" (half the volume of "The Cube").

What could be the dimensions of the three storage units? How might "The Double" and "The Half" be oriented to cover the most floor space? The most height?

Name: ______________________ Date: __________

GRADE 5 / MISSION 5 / LESSON 6

Exit Ticket

1. The image below represents three planters that are filled with soil. Find the total volume of soil in the three planters. Planter A is 14 inches by 3 inches by 4 inches. Planter B is 9 inches by 3 inches by 3 inches.

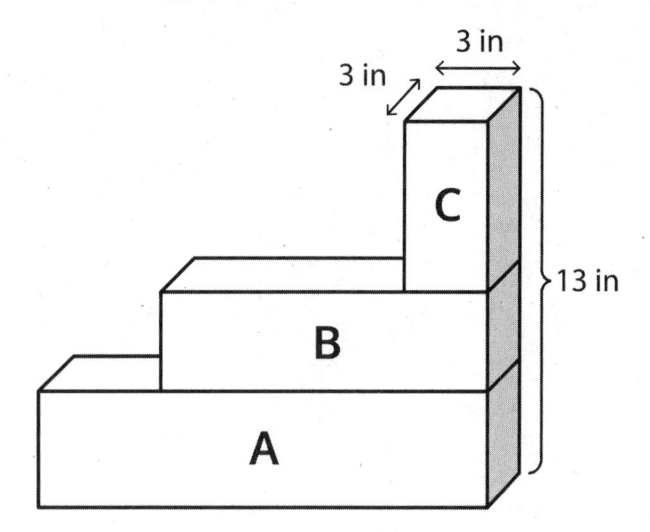

Lesson 7

Name: ______________________ Date: __________

GRADE 5 / MISSION 5 / LESSON 7

Exit Ticket

1. A storage shed is a rectangular prism and has dimensions of 6 meters by 5 meters by 12 meters. If Jean were to double these dimensions, she believes she would only double the volume.

   Is she correct? Explain why or why not. Include a drawing in your explanation.

PROBLEM SET

Geoffrey builds rectangular planters.

1. Geoffrey's first planter is 8 feet long and 2 feet wide. The container is filled with soil to a height of 3 feet in the planter. What is the volume of soil in the planter? Explain your work using a diagram.

2. Geoffrey wants to grow some tomatoes in four large planters. He wants each planter to have a volume of 320 cubic feet, but he wants them all to be different. Show four different ways Geoffrey can make these planters, and draw diagrams with the planters' measurements on them.

Planter A	Planter B
Planter C	Planter D

3. Geoffrey wants to make one planter that extends from the ground to just below his back window. The window starts 3 feet off the ground. If he wants the planter to hold 36 cubic feet of soil, name one way he could build the planter so it is not taller than 3 feet. Explain how you know.

4. After all of this gardening work, Geoffrey decides he needs a new shed to replace the old one. His current shed is a rectangular prism that measures 6 feet long by 5 feet wide by 8 feet high. He realizes he needs a shed with 480 cubic feet of storage.

   a. Will he achieve his goal if he doubles each dimension? Why or why not?

   b. If he wants to keep the height the same, what could the other dimensions be for him to get the volume he wants?

   c. If he uses the dimensions in part (b), what could be the area of the new shed's floor?

Lesson 8

Name: ______________________ **Date:** __________

GRADE 5 / MISSION 5 / LESSON 8

Exit Ticket

1. Sketch a rectangular prism that has a volume of 36 cubic cm. Label the dimensions of each side on the prism. Fill in the blanks that follow.

   Height: ________ cm

   Length: ________ cm

   Width: ________ cm

   Volume: _______ cubic cm

PROBLEM SET

Using the box patterns, construct a sculpture containing at least 5, but not more than 7, rectangular prisms that meets the following requirements in the table below.

1.	My sculpture has 5 to 7 rectangular prisms. Number of prisms: ______	
2.	Each prism is labeled with a letter, dimensions, and volume.	
	Prism A ______ by ______ by ______ Volume = ______ Prism B ______ by ______ by ______ Volume = ______ Prism C ______ by ______ by ______ Volume = ______ Prism D ______ by ______ by ______ Volume = ______ Prism E ______ by ______ by ______ Volume = ______ Prism ____ ______ by ______ by ______ Volume = ______ Prism ____ ______ by ______ by ______ Volume = ______	
3.	Prism D has $\frac{1}{2}$ the volume of Prism ____.	Prism D Volume = ______ Prism ____ Volume = ______
4.	Prism E has $\frac{1}{3}$ the volume of Prism ____.	Prism E Volume = ______ Prism ____ Volume = ______
5.	The total volume of all the prisms is 1,000 cubic centimeters or less.	Total volume: ______ Show calculations:

Lesson 9

Word Problem

The chart below shows the dimensions of various rectangular packing boxes. If possible, answer the following without calculating the volume.

a. Which box will provide the greatest volume?

b. Which box has a volume that is equal to the volume of the book box? How do you know?

c. Which box is $\frac{1}{3}$ the volume of the lamp box?

Box Type	Dimensions (l × w × h)
Book Box	12 in × 12 in × 12 in
Picture Box	36 in × 12 in × 36 in
Lamp Box	12 in × 9 in × 48 in
The Flat	12 in × 6 in × 24 in

Name: ______________________ **Date:** __________

GRADE 5 / MISSION 5 / LESSON 9

Exit Ticket

A student designed this sculpture. Using the dimensions on the sculpture, find the dimensions of each rectangular prism. Then, calculate the volume of each prism.

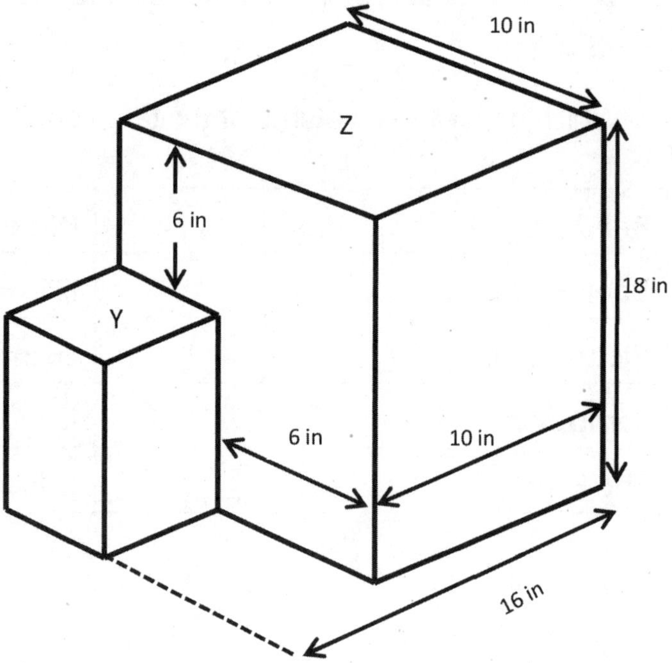

a. Rectangular Prism Y

Height: ________ inches

Length: ________ inches

Width: ________ inches

Volume: ________ cubic inches

b. Rectangular Prism Z

Height: ________ inches

Length: ________inches

Width: ________ inches

Volume: ________ cubic inches

10 in
Z
6 in
18 in
Y
6 in
10 in
16 in

c. Find the total volume of the sculpture. Label the answer.

PROBLEM SET

I reviewed project number ______________.

Use the rubric below to evaluate your friend's project. Ask questions and measure the parts to determine whether your friend has all the required elements. Respond to the prompt in italics in the third column. The final column can be used to write something you find interesting about that element if you like.

Space is provided beneath the rubric for your calculations.

	Requirement	Element Present? (✓)	Specifics of Element	Notes
1.	The sculpture has 5 to 7 prisms.		*# of prisms:*	
2.	All prisms are labeled with a letter.		*Write letters used:*	
3.	All prisms have correct dimensions with units written on the top.		*List any prisms with incorrect dimensions or units:*	
4.	All prisms have correct volume with units written on the top.		*List any prism with incorrect dimensions or units:*	
5.	Prism D has $\frac{1}{2}$ the volume of another prism.		*Record on next page:*	
6.	Prism E has $\frac{1}{3}$ the volume of another prism.		*Record on next page:*	
7.	The total volume of all the parts together is 1,000 cubic units or less.		*Total volume:*	

Calculations:

8. Measure the dimensions of each prism. Calculate the volume of each prism and the total volume. Record that information in the table below. If your measurements or volume differ from those listed on the project, put a star by the prism label in the table below, and record on the rubric.

Prism	Dimensions	Volume
A	_______ by _______ by _______	
B	_______ by _______ by _______	
C	_______ by _______ by _______	
D	_______ by _______ by _______	
E	_______ by _______ by _______	
	_______ by _______ by _______	
	_______ by _______ by _______	

9. Prism D's volume is $\frac{1}{2}$ that of Prism __________.

Show calculations below.

10. Prism E's volume is $\frac{1}{3}$ that of Prism __________.

Show calculations below.

11. Total volume of sculpture: __________.

Show calculations below.

EVALUATION RUBRIC (LESSON 8 CONCEPT EXPLORATION TEMPLATE 5)

CATEGORY	4	3	2	1	Subtotal
Completeness of Personal Project and Classmate Evaluation	All components of the project are present and correct, and a detailed evaluation of a classmate's project has been completed.	Project is missing 1 component, and a detailed evaluation of a classmate's project has been completed.	Project is missing 2 components, and an evaluation of a classmate's project has been completed.	Project is missing 3 or more components, and an evaluation of a classmate's project has been completed.	(× 4) ______/16
Accuracy of Calculations	Volume calculations for all prisms are correct.	Volume calculations include 1 error.	Volume calculations include 2–3 errors.	Volume calculations include 4 or more errors.	(× 5) ______/20
Neatness and Use of Color	All elements of the project are carefully and colorfully constructed.	Some elements of the project are carefully and colorfully constructed.	Project lacks color or is not carefully constructed.	Project lacks color and is not carefully constructed.	(× 2) ______/4
					TOTAL: ______/40

Lesson 10

Word Problem

Heidi and Andrew designed two raised flowerbeds for their garden. Heidi's flowerbed was 5 feet long by 3 feet wide, and Andrew's flowerbed was the same length but twice as wide. Calculate how many cubic feet of soil they need to buy to have soil to a depth of 2 feet in both flowerbeds.

Name: ______________________________ **Date:** __________

GRADE 5 / MISSION 5 / LESSON 10

Exit Ticket

1. Emma tiled a rectangle and then sketched her work.

   Fill in the missing information, and multiply to find the area.

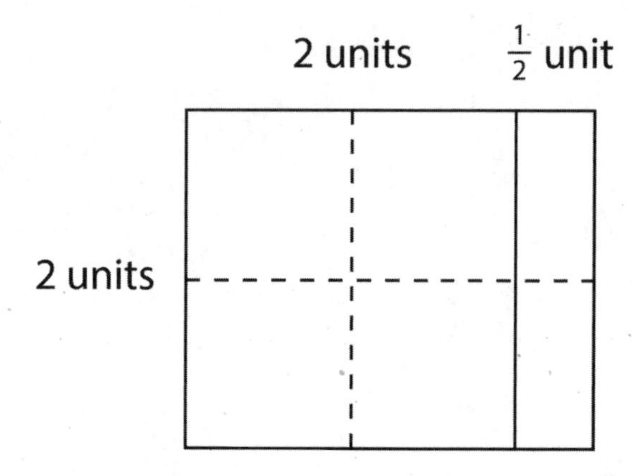

Emma's Rectangle:

________ units long

________ units wide

Area = ________ units²

PROBLEM SET

Sketch the rectangles and your tiling. Write the dimensions and the units you counted in the blanks. Then, use multiplication to confirm the area. Show your work. We will do Rectangles A and B together.

1. **Rectangle A:**

Rectangle A is

_____ units long _____ units wide

Area = _____ $units^2$

2. **Rectangle B:**

Rectangle B is

_____ units long _____ units wide

Area = _____ $units^2$

3. **Rectangle C:**

Rectangle C is

_____ units long _____ units wide

Area = _____ $units^2$

4. **Rectangle D:**

Rectangle D is

_____ units long _____ units wide

Area = _____ $units^2$

5. **Rectangle E:**

Rectangle E is

_____ units long _____ units wide

Area = _____ $units^2$

6. The rectangle to the right is composed of squares that measure $2\frac{1}{4}$ inches on each side. What is its area in square inches? Explain your thinking using pictures and numbers.

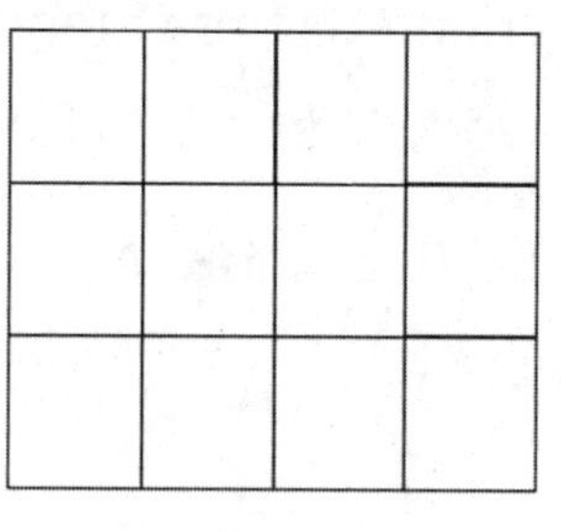

7. A rectangle has a perimeter of $35\frac{1}{2}$ feet. If the length is 12 feet, what is the area of the rectangle?

Lesson 11

Word Problem

Mrs. Golden wants to cover her 6.5-foot by 4-foot bulletin board with silver paper that comes in 1-foot squares. How many squares does Mrs. Golden need to cover her bulletin board? Will there be any fractional pieces of silver paper left over? Explain why or why not. Draw a sketch to show your thinking.

Name: ______________________ **Date:** __________

GRADE 5 / MISSION 5 / LESSON 11

Exit Ticket

1. To find the area, Andrea tiled a rectangle and sketched her answer.

   Sketch Andrea's rectangle, and find the area. Show your multiplication work.

   Rectangle is $2\frac{1}{2}$ units × $2\frac{1}{2}$ units

   Area = ______________

PROBLEM SET

Draw the rectangle and your tiling. Write the dimensions and the units you counted in the blanks. Then, use multiplication to confirm the area. Show your work.

1. **Rectangle A:**

Rectangle A is

_______ units long _______ units wide

Area = _______ $units^2$

2. **Rectangle B:**

Rectangle B is

_______ units long _______ units wide

Area = _______ $units^2$

3. **Rectangle C:**

Rectangle C is

_______ units long _______ units wide

Area = _______ $units^2$

4. **Rectangle D:**

Rectangle D is

_______ units long _______ units wide

Area = _______ $units^2$

5. Colleen and Caroline each built a rectangle out of square tiles placed in 3 rows of 5. Colleen used tiles that measured $1\frac{2}{3}$ cm in length. Caroline used tiles that measured $3\frac{1}{3}$ cm in length.

   a. Draw the girls' rectangles, and label the lengths and widths of each.

   b. What are the areas of the rectangles in square centimeters?

   c. Compare the areas of the rectangles.

6. A square has a perimeter of 51 inches. What is the area of the square?

Lesson 12

Word Problem

Margo is designing a label. The dimensions of the label are $3\frac{1}{2}$ inches by $1\frac{1}{4}$ inches. What is the area of the label? Use the RDW process.

Name: ______________________________ Date: __________

GRADE 5 / MISSION 5 / LESSON 12

Exit Ticket

1. Measure the rectangle to the nearest $\frac{1}{4}$ inch with your ruler, and label the dimensions. Find the area.

PROBLEM SET

1. Measure each rectangle to the nearest $\frac{1}{4}$ inch with your ruler, and label the dimensions. Use the area model to find each area.

a.

b.

c.

d.

e.

f.

2. Find the area of rectangles with the following dimensions. Explain your thinking using the area model.

   a. 1 ft × $1\frac{1}{2}$ ft

   b. $1\frac{1}{2}$ yd × $1\frac{1}{2}$ yd

   c. $2\frac{1}{2}$ yd × $1\frac{3}{16}$ yd

3. Hanley is putting carpet in her house. She wants to carpet her living room, which measures 15 ft × $12\frac{1}{3}$ ft. She also wants to carpet her dining room, which is $10\frac{1}{4}$ ft × $10\frac{1}{3}$ ft. How many square feet of carpet will she need to cover both rooms?

4. Fred cut a $9\frac{3}{4}$-inch square of construction paper for an art project. He cut a square from the edge of the big rectangle whose sides measured $3\frac{1}{4}$ inches. (See the picture below.)

   a. What is the area of the smaller square that Fred cut out?

   b. What is the area of the remaining paper?

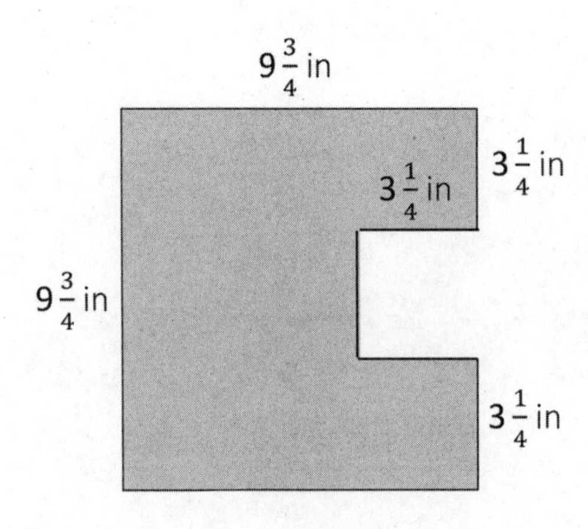

Lesson 13

Word Problem

The Colliers want to put new flooring in a $6\frac{1}{2}$-foot by $7\frac{1}{3}$-foot bathroom. The tiles they want come in 12-inch squares. What is the area of the bathroom floor? If the tiles cost $3.25 per square foot, how much will they spend on the flooring?

Name: ______________________ Date: __________

GRADE 5 / MISSION 5 / LESSON 13

Exit Ticket

Find the area of the following rectangles. Draw an area model if it helps you.

1. $\frac{7}{2}$ mm × $\frac{14}{5}$ mm

2. $5\frac{7}{8}$ km × $\frac{18}{4}$ km

Lesson 14

Name: ______________________ Date: ____________

GRADE 5 / MISSION 5 / LESSON 14

Exit Ticket

1. Mr. Klimek made his wife a rectangular vegetable garden. The width is $5\frac{3}{4}$ ft, and the length is $9\frac{4}{5}$ ft.

   What is the area of the garden?

PROBLEM SET

1. George decided to paint a wall with two windows. Both windows are $3\frac{1}{2}$-ft by $4\frac{1}{2}$-ft rectangles. Find the area the paint needs to cover.

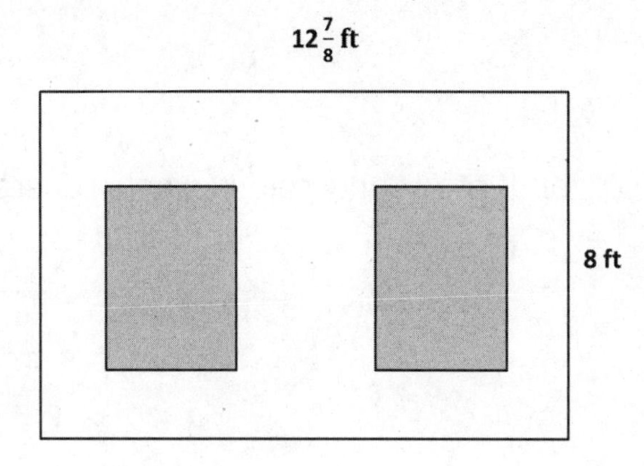

2. Joe uses square tiles, some of which he cuts in half, to make the figure below. If each square tile has a side length of $2\frac{1}{2}$ inches, what is the total area of the figure?

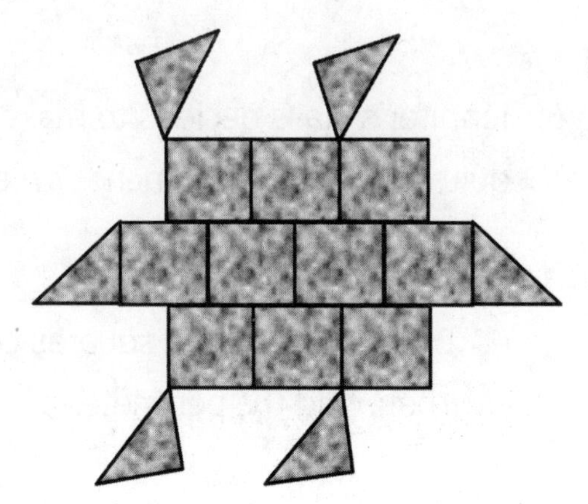

3. All-In-One Carpets is installing carpeting in three rooms. How many square feet of carpet are needed to carpet all three rooms?

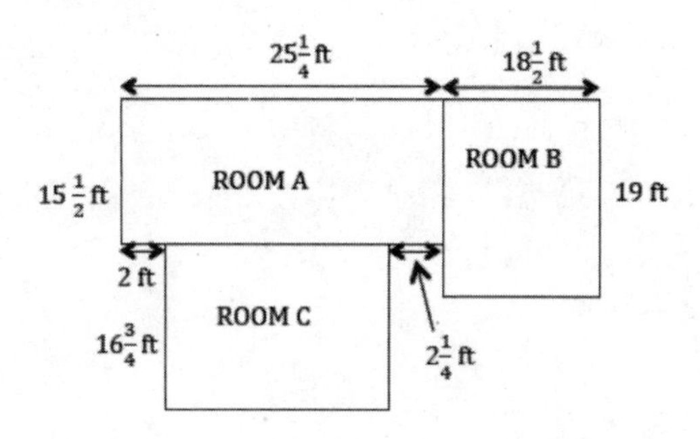

4. Mr. Johnson needs to buy sod for his front lawn.

   a. If the lawn measures $36\frac{2}{3}$ ft by $45\frac{1}{6}$ ft, how many square feet of sod will he need?

   b. If sod is only sold in whole square feet, how much will Mr. Johnson have to pay?

Sod Prices	
Area	**Price per Square Foot**
First 1,000 sq ft	\$0.27
Next 500 sq ft	\$0.22
Additional square feet	\$0.19

5. Jennifer's class decides to make a quilt. Each of the 24 students will make a quilt square that is 8 inches on each side. When they sew the quilt together, every edge of each quilt square will lose $\frac{3}{4}$ of an inch.

   a. Draw one way the squares could be arranged to make a rectangular quilt. Then, find the perimeter of your arrangement.

   b. Find the area of the quilt.

Lesson 15

Name: ________________________ Date: ____________

GRADE 5 / MISSION 5 / LESSON 15

Exit Ticket

1. Wheat grass is grown in planters that are $3\frac{1}{2}$ inches by $1\frac{3}{4}$ inches. If there is a 6×6 array of these planters with no space between them, what is the area covered by the planters?

PROBLEM SET

1. The length of a flowerbed is 4 times as long as its width. If the width is $\frac{3}{8}$ meter, what is the area?

2. Mrs. Johnson grows herbs in square plots. Her basil plot measures $\frac{5}{8}$ yd on each side.

   a. Find the total area of the basil plot.

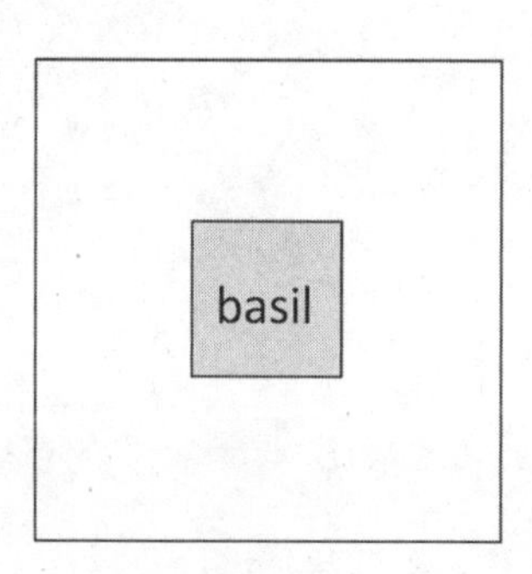

   b. Mrs. Johnson puts a fence around the basil. If the fence is 2 ft from the edge of the garden on each side, what is the perimeter of the fence in feet?

   c. What is the total area, in square feet, that the fence encloses?

3. Janet bought 5 yards of fabric $2\frac{1}{4}$-feet wide to make curtains. She used $\frac{1}{3}$ of the fabric to make a long set of curtains and the rest to make 4 short sets.

   a. Find the area of the fabric she used for the long set of curtains.

   b. Find the area of the fabric she used for each of the short sets.

4. Some wire is used to make 3 rectangles: A, B, and C. Rectangle B's dimensions are $\frac{3}{5}$ cm larger than Rectangle A's dimensions, and Rectangle C's dimensions are $\frac{3}{5}$ cm larger than Rectangle B's dimensions. Rectangle A is 2 cm by $3\frac{1}{5}$ cm.

   a. What is the total area of all three rectangles?

   b. If a 40-cm coil of wire was used to form the rectangles, how much wire is left?

Lesson 16

Word Problem

Kathy spent 3 fifths of her money on a necklace and 2 thirds of the remainder on a bracelet. If the bracelet cost $17, how much money did she have at first?

Name: ______________________ Date: ____________

GRADE 5 / MISSION 5 / LESSON 16

Exit Ticket

1. Use a ruler and a set square to draw a trapezoid.

2. What attribute must be present for a quadrilateral to also be a trapezoid?

QUADRILATERAL HIERARCHY (CONCEPT EXPLORATION TEMPLATE 2)

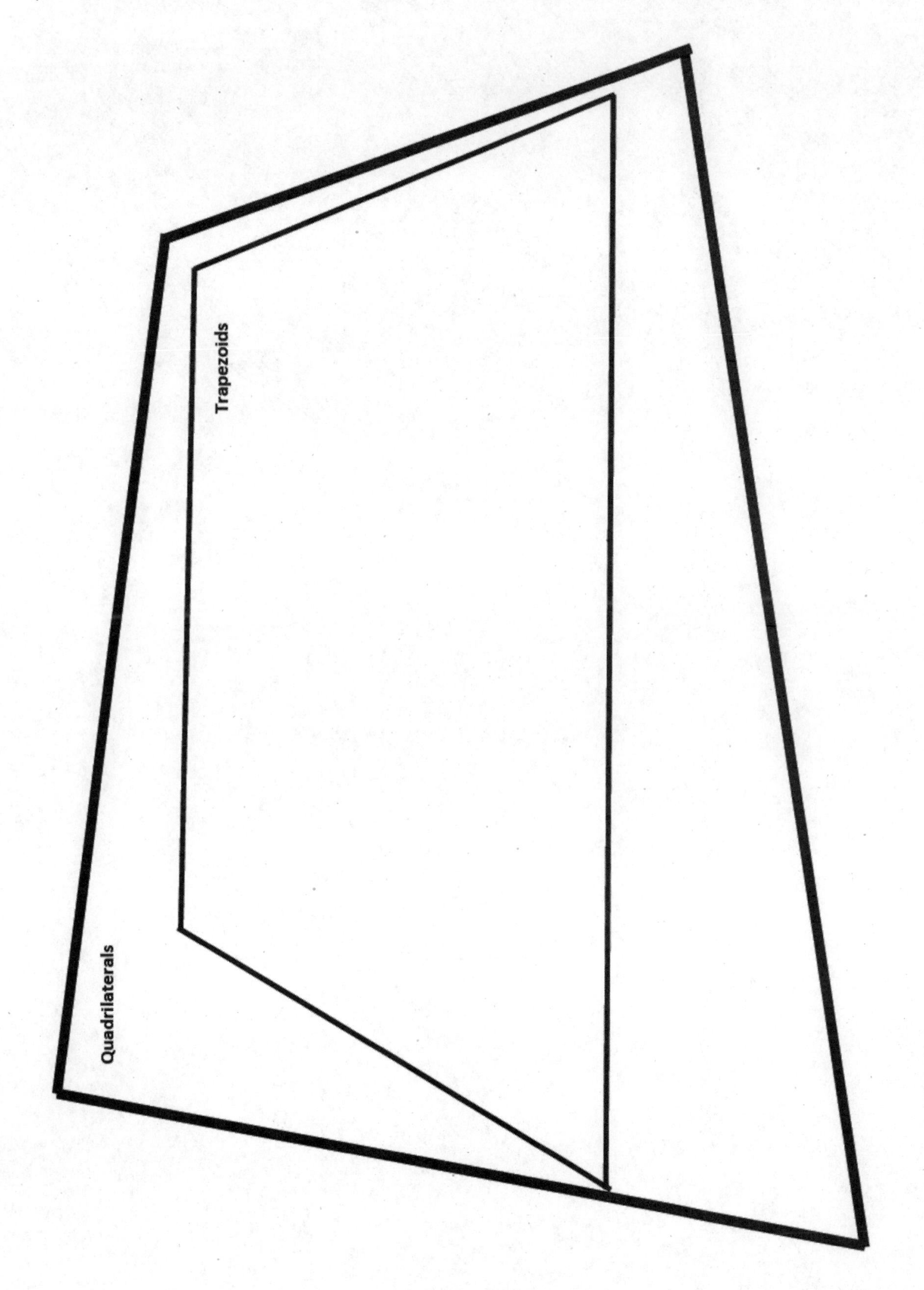

Lesson 17

Word Problem

Ava drew the quadrilateral pictured and called it a trapezoid. Adam said Ava is wrong. Explain to your partner how a set square can be used to determine who is correct. Support your answer using the properties of trapezoids.

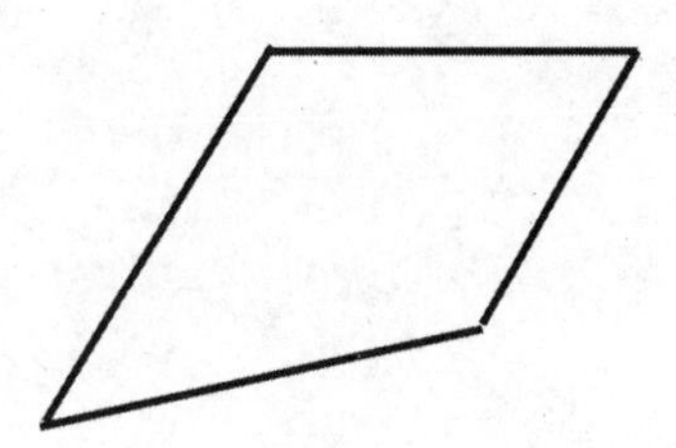

Name: ______________________ **Date:** __________

GRADE 5 / MISSION 5 / LESSON 17

Exit Ticket

1. Draw a parallelogram.

2. When is a trapezoid also called a parallelogram?

QUADRILATERAL HIERARCHY WITH PARALLELOGRAM (CONCEPT EXPLORATION TEMPLATE 1)

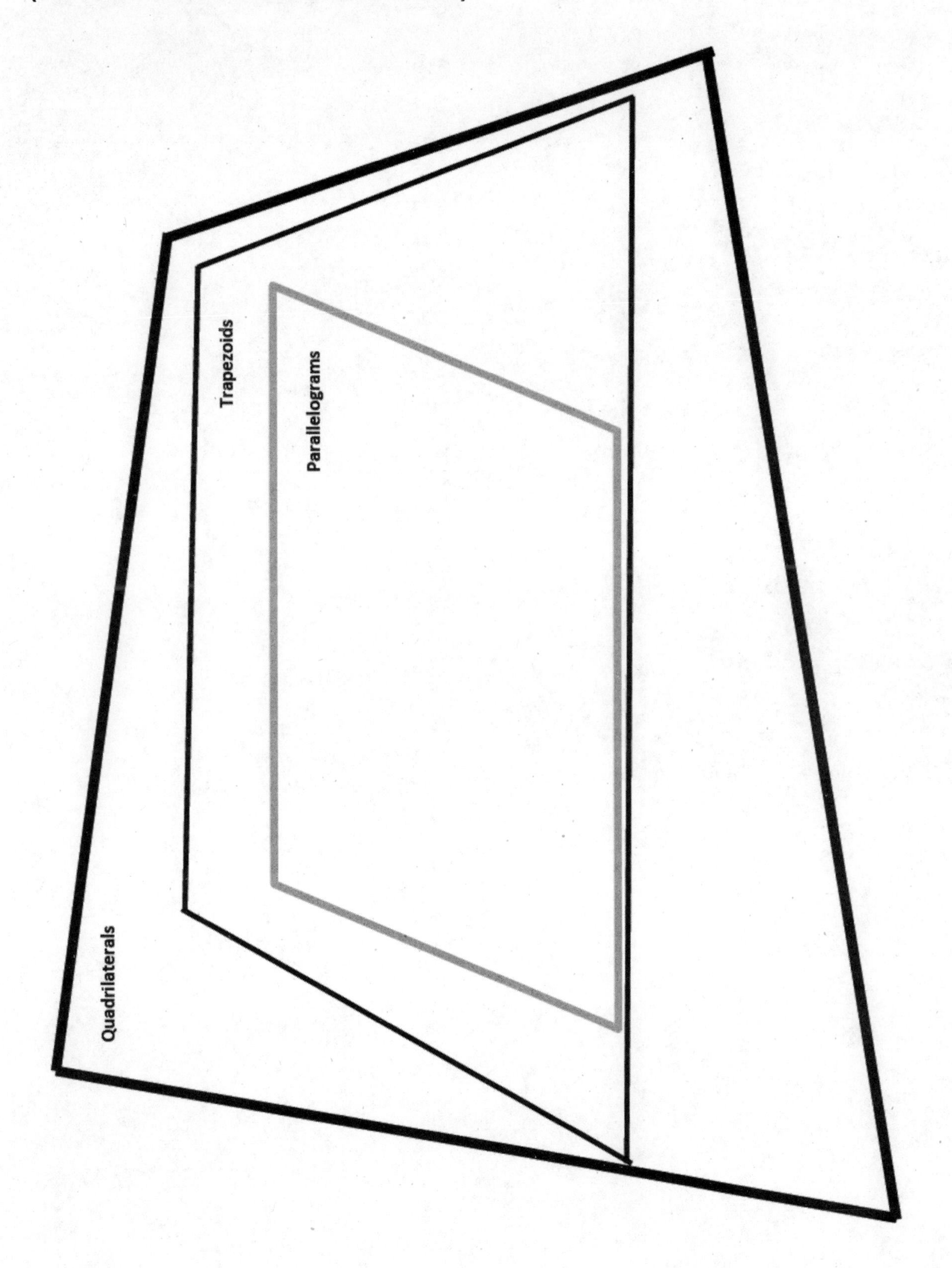

Lesson 18

Word Problem

How many 2-inch cubes are needed to build a rectangular prism that measures 10 inches by 14 inches by 6 inches?

Name: ______________________ Date: __________

GRADE 5 / MISSION 5 / LESSON 18

Exit Ticket

1. Draw a rhombus.

2. Draw a rectangle.

QUADRILATERAL HIERARCHY WITH SQUARE (CONCEPT EXPLORATION TEMPLATE 1)

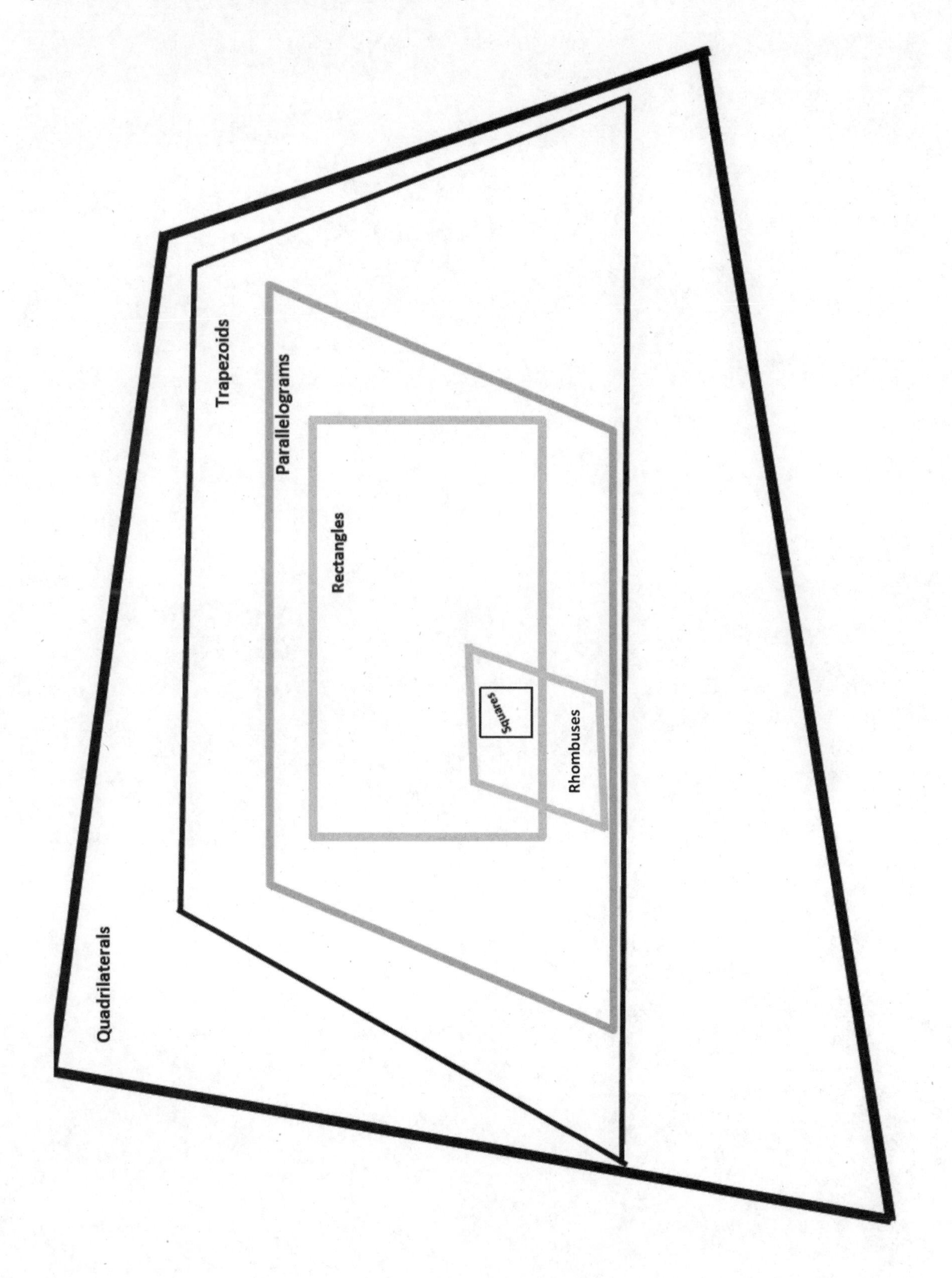

Lesson 19

Word Problem

The teacher asked her class to draw parallelograms that are rectangles. Kylie drew Figure 1, and Zach drew Figure 2. Zach agrees that Kylie has drawn a parallelogram but says that it is not a rectangle. Is he correct? Use properties to justify your answer.

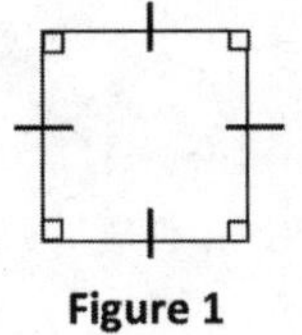

Figure 1

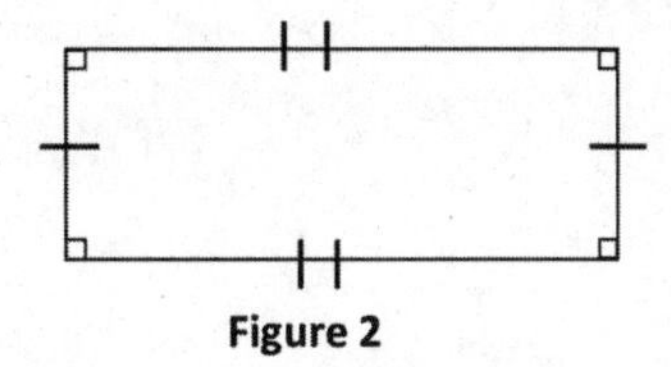

Figure 2

Name: ______________________________ Date: ____________

GRADE 5 / MISSION 5 / LESSON 19

Exit Ticket

1. Draw a square.

2. List the property that must be present to call a rectangle a square.

QUADRILATERAL HIERARCHY WITH KITE (CONCEPT EXPLPORATION TEMPLATE 1)

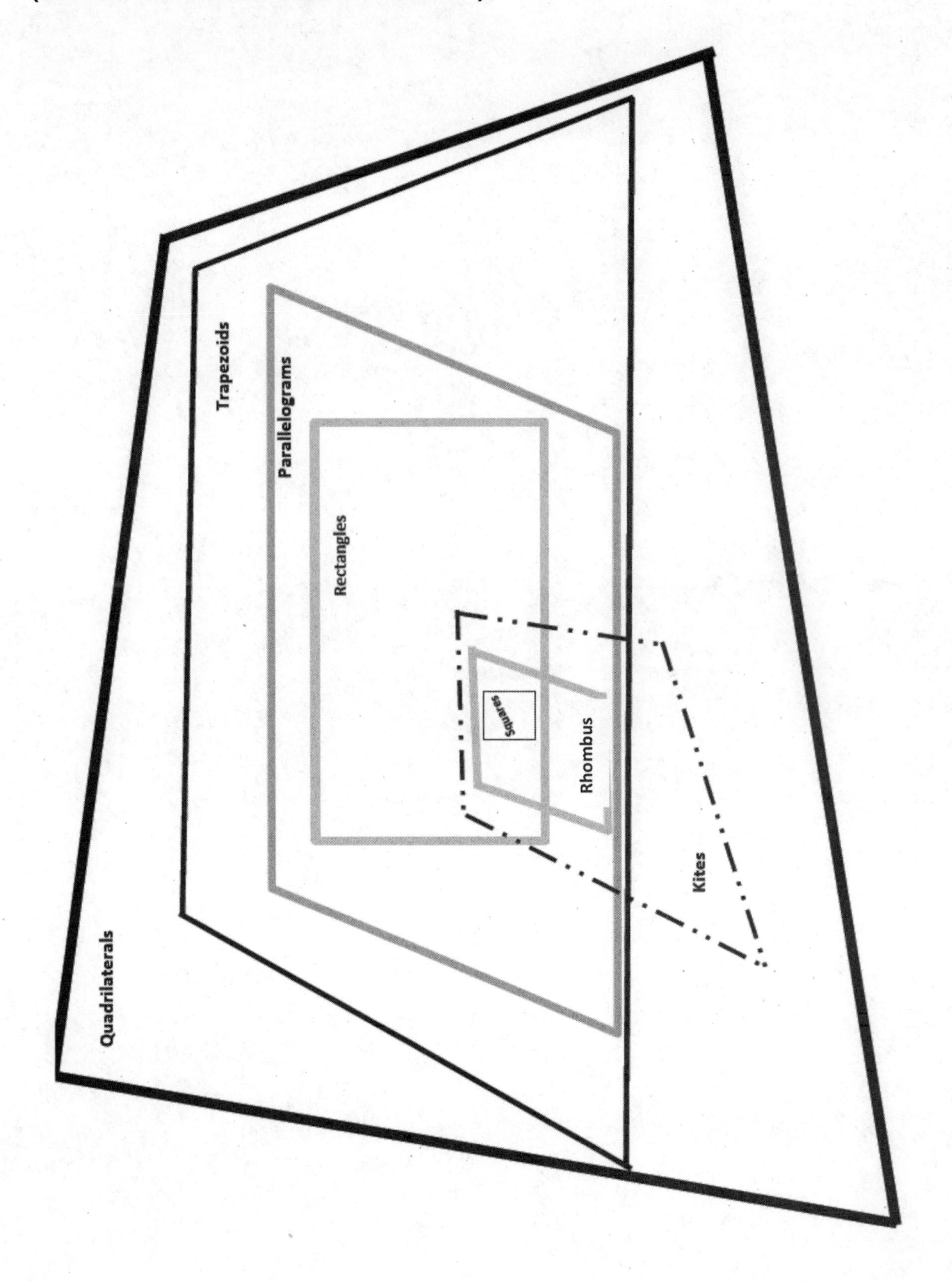

Lesson 20

Word Problem

Nita buys a rug that is $10\frac{3}{4}$ feet × $12\frac{1}{2}$ feet. What is the area of the rug? Show your thinking with an area model and a multiplication sentence.

Name: ______________________ Date: __________

GRADE 5 / MISSION 5 / LESSON 20

Exit Ticket

1. Use your tools to draw a square in the space below. Then, fill in the blanks with an attribute. There is more than one answer to some of these.

   a. Because a square is a kite, it must have

   ____.

   b. Because a square is a rhombus, it must have

   ____.

   c. Because a square is a rectangle, it must have

   ____.

   d. Because a square is a parallelogram, it must have

   ____.

   e. Because a square is a trapezoid, it must have

   ____.

   f. Because a square is a quadrilateral, it must have

   ____.

QUADRILATERAL HIERARCHY WITH KITE (CONCEPT EXPLPORATION TEMPLATE 1)

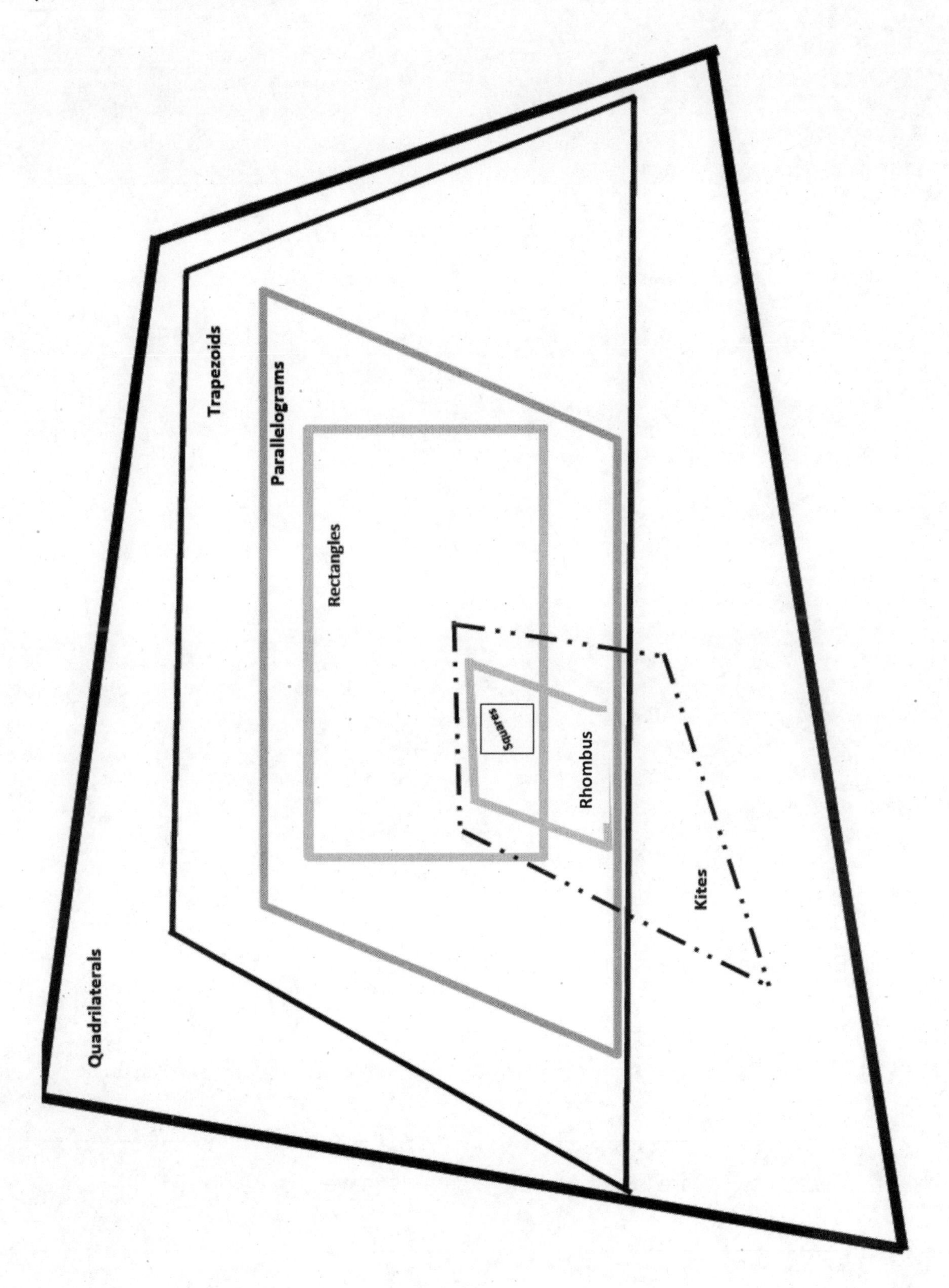

Lesson 21

Name: ______________________ Date: __________

GRADE 5 / MISSION 5 / LESSON 21

Exit Ticket

1. Use the word bank to fill in the blanks.

trapezoids **parallelograms**

All ______________ are ______________,

but not all ______________ are ______________.

2. Use the word bank to fill in the blanks.

squares **rhombuses**

All ______________ are ______________,

but not all ______________ are ______________.

PROBLEM SET

1. Write the number on your task card and a summary of the task in the blank. Then, draw the figure in the box. Label your figure with as many names as you can. Circle the most specific name.

Task #___: ______________	Task #___: ______________
Task #___: ______________	Task #___: ______________
Task #___: ______________	Task #___: ______________

2. John says that because rhombuses do not have perpendicular sides, they cannot be rectangles. Explain his error in thinking.

3. Jack says that because kites do not have parallel sides, a square is not a kite. Explain his error in thinking.

Grade 5

Mission 6

The Coordinate Plane

Lesson 1

Word Problem

A landscaper is planting some marigolds in a row. The row is 2 yards long. The flowers must be spaced $\frac{1}{3}$ yard apart so that they will have proper room to grow. The landscaper plants the first flower at 0.

Place points on the number line to show where the landscaper should place the other flowers. How many marigolds will fit in this row?

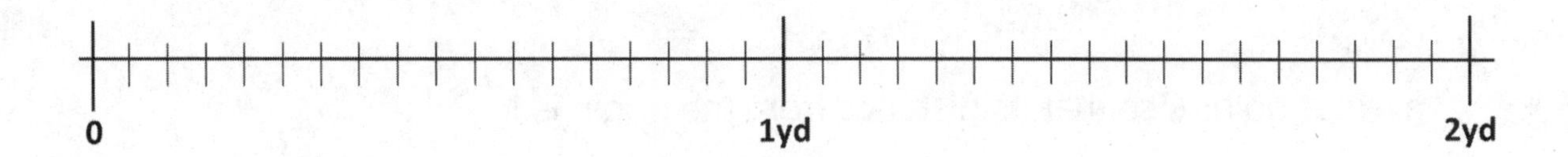

Name: ______________________ Date: __________

GRADE 5 / MISSION 6 / LESSON 1

Exit Ticket

1. Use number line l to answer the questions.

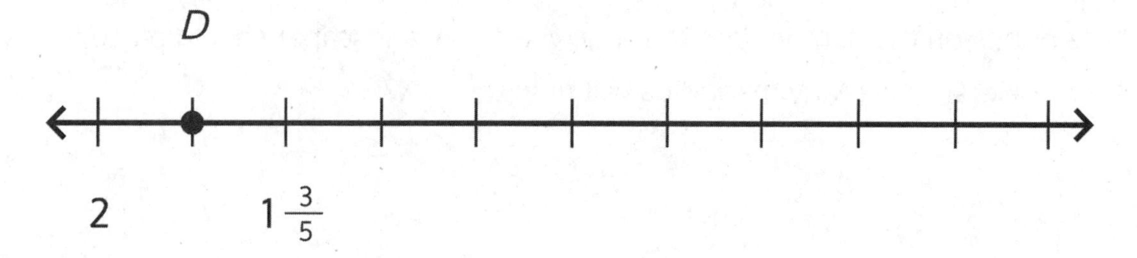

a. Plot point C so that its distance from the origin is 1.

b. Plot point E $\frac{4}{5}$ closer to the origin than C. What is its coordinate?

____

c. Plot a point at the midpoint of C and E. Label it H.

Lesson 2

Word Problem

The picture shows an intersection in Stony Brook Village.

a. The town wants to construct two new roads, Elm Street and King Street. Elm Street will intersect Lower Sheep Pasture Road, run parallel to Main Street, and be perpendicular to Stony Brook Road. Sketch Elm Street.

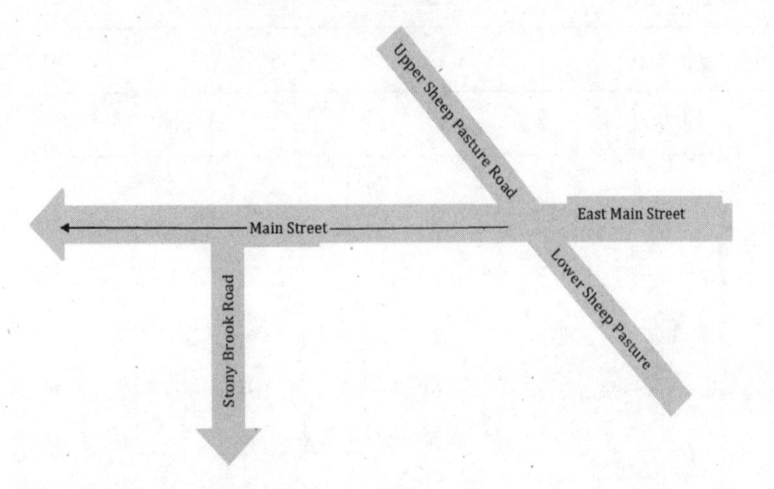

b. King Street will be perpendicular to Main Street and begin at the intersection of Upper Sheep Pasture Road and East Main Street. Sketch King Street.

Name: ______________________ Date: __________

GRADE 5 / MISSION 6 / LESSON 2

Exit Ticket

1. Name the coordinates of the shapes below.

Shape	*x*-coordinate	*y*-coordinate
Star		
Arrow		
Heart		

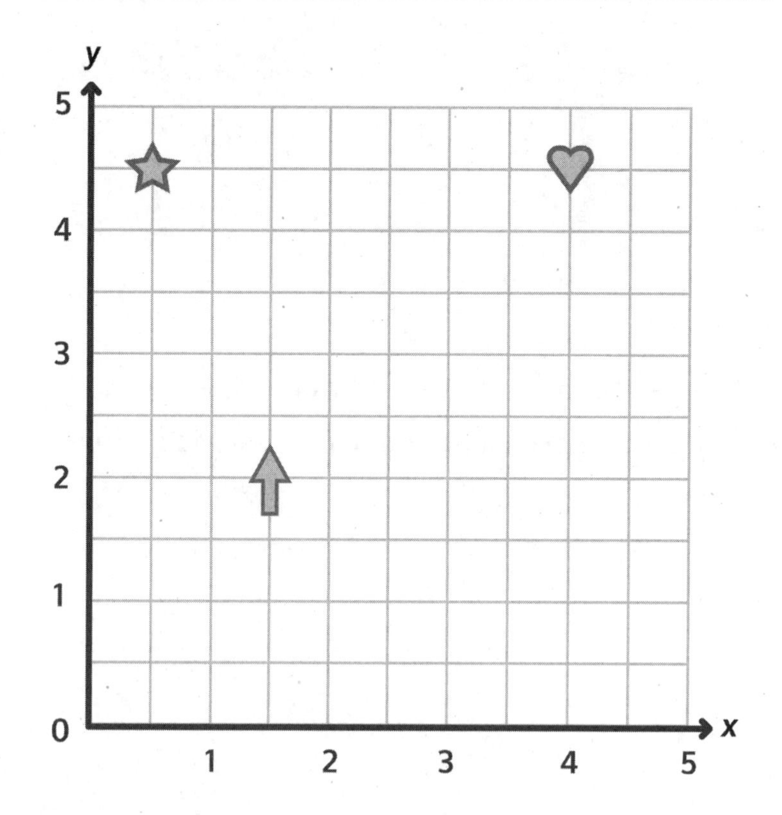

2. Plot a square at $(3, 3\frac{1}{2})$.

3. Plot a triangle at $(4\frac{1}{2}, 1)$.

COORDINATE PLANE (CONCEPT EXPLORATION TEMPLATE)

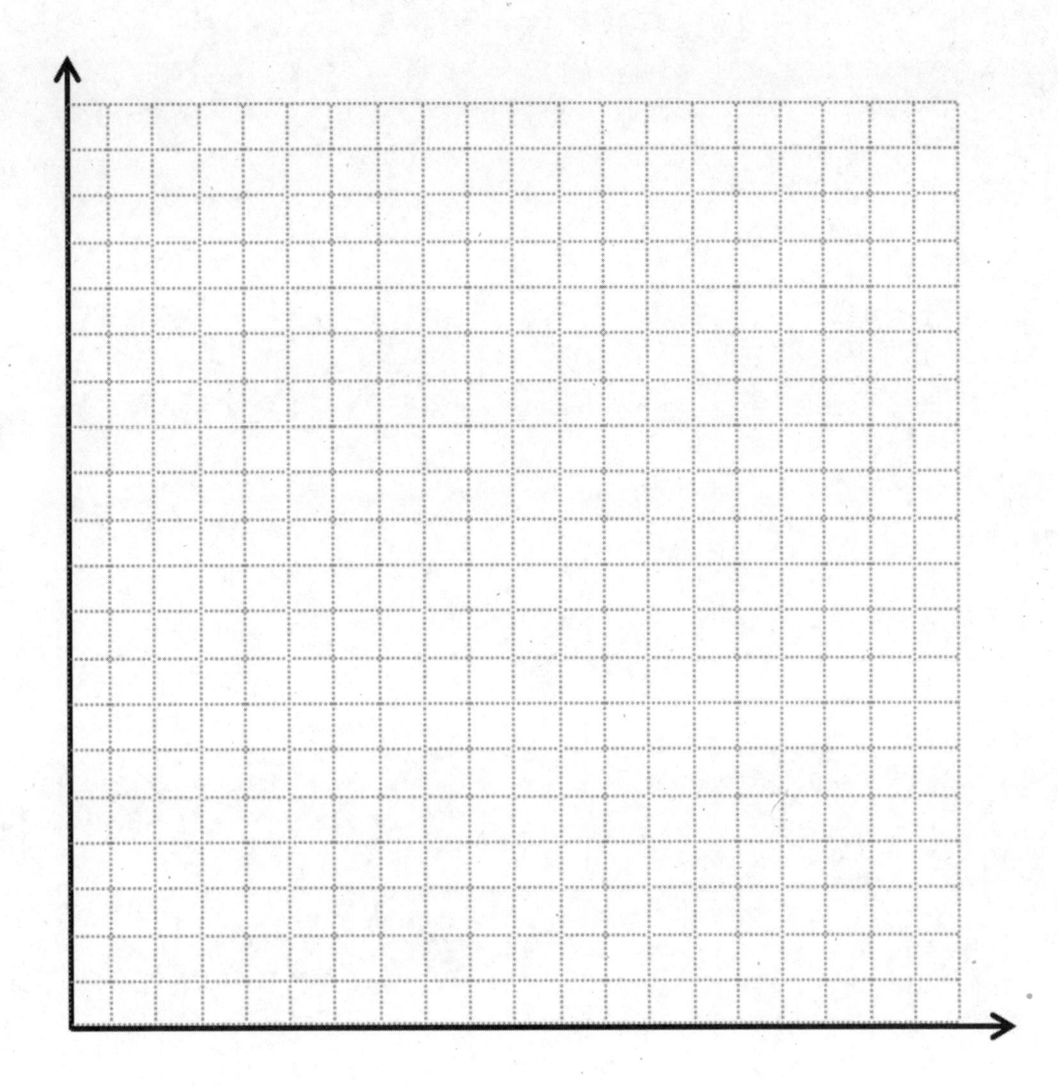

Lesson 3

Word Problem

The captain of a ship has a chart to help him navigate through the islands. He must follow points that show the deepest part of the channel. List the coordinates the captain needs to follow in the order he will encounter them.

1. (________, ________) 2. (________, ________)

3. (________, ________) 4. (________, ________)

5. (________, ________) 6. (________, ________)

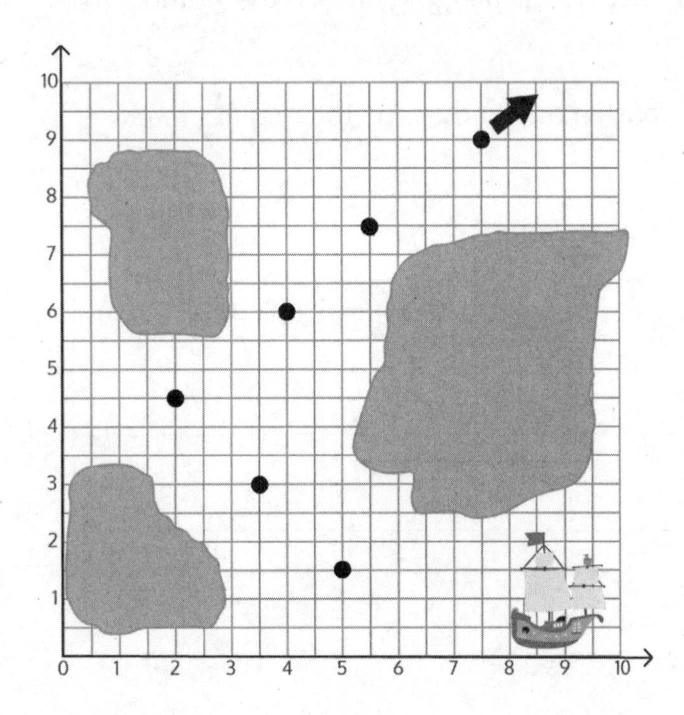

Name: ______________________ **Date:** __________

GRADE 5 / MISSION 6 / LESSON 3

Exit Ticket

1. Use a ruler on the grid below to construct the axes for a coordinate plane. The *x*-axis should intersect points *L* and *M*. Construct the *y*-axis so that it contains points *K* and *L*. Label each axis.

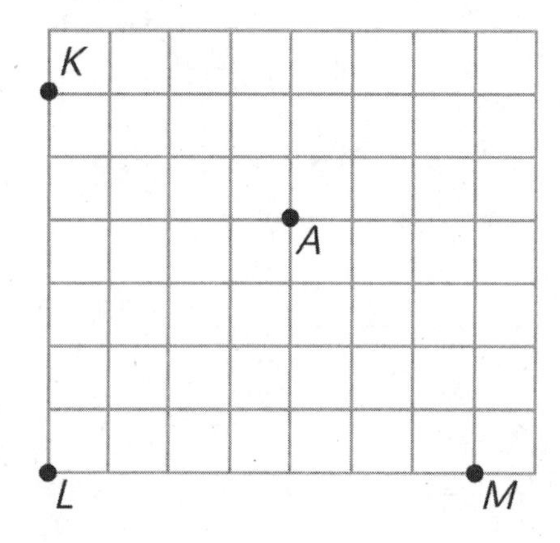

a. Place a hash mark on each grid line on the *x*- and *y*-axis.

b. Label each hash mark so that *A* is located at (1 , 1).

c. Plot the following points:

Point	*x*-coordinate	*y*-coordinate
B	$\frac{1}{4}$	0
C	$1\frac{1}{4}$	$\frac{3}{4}$

UNLABELED COORDINATE PLANE (CONCEPT EXPLORATION TEMPLATE)

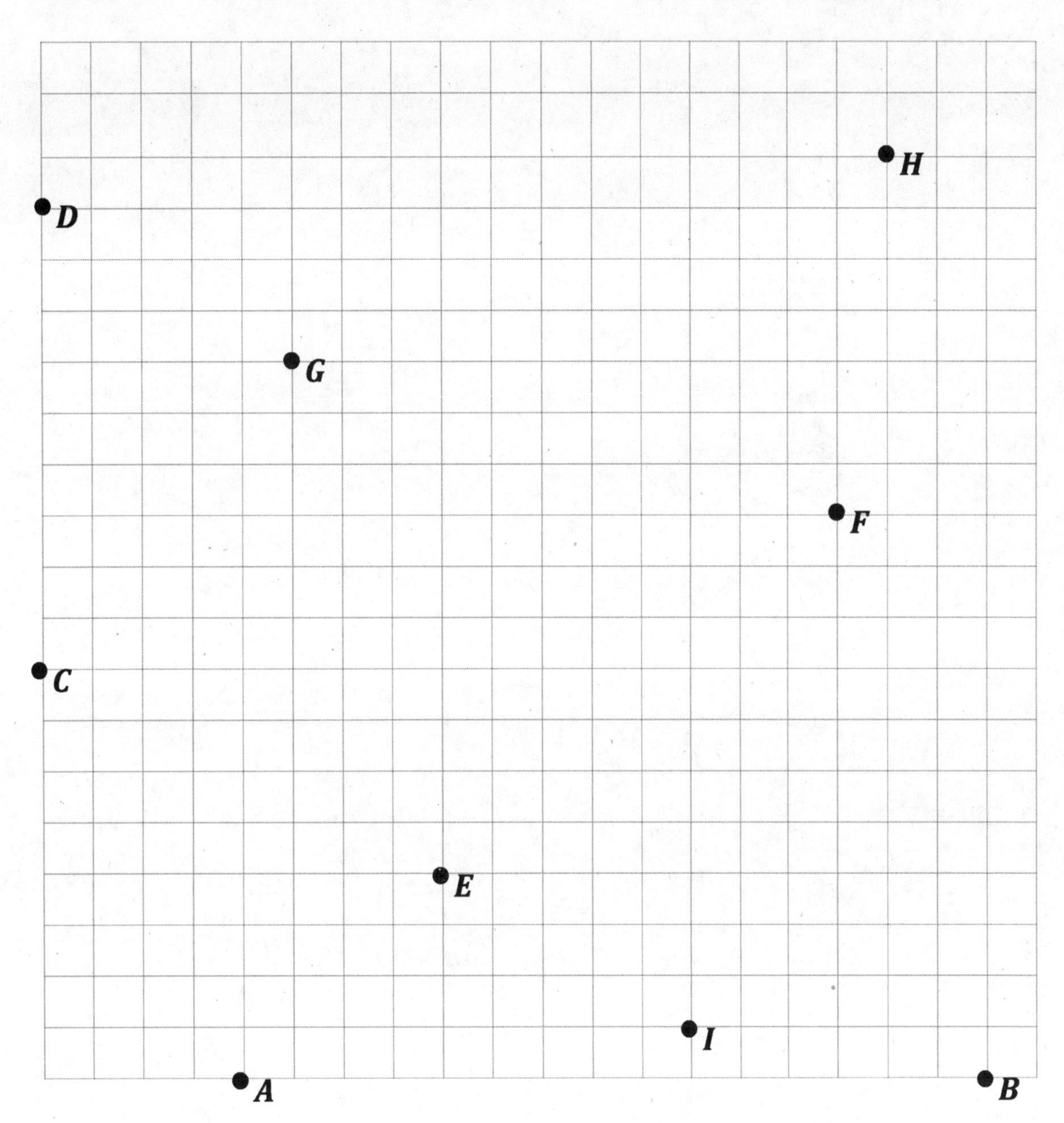

Lesson 4

Word Problem

Violet and Magnolia are shopping for boxes to organize the materials for their design company. Magnolia wants to get small boxes, which measure 16 in × 10 in × 7 in. Violet wants to get large boxes, which measure 32 in × 20 in × 14 in. How many small boxes will equal the volume of four large boxes?

Name: ______________________ Date: __________

GRADE 5 / MISSION 6 / LESSON 4

Exit Ticket

a. Fatima and Rihana are playing Battleship. They labeled their axes using just whole numbers. Fatima's first guess is (2, 2). Rihana says, "Hit!" Give the coordinates of four points that Fatima might guess next.

b. Rihana says, "Hit!" for the points directly above and below (2, 2). What are the coordinates that Fatima guessed?

COORDINATE GRID (FLUENCY TEMPLATE)

a.

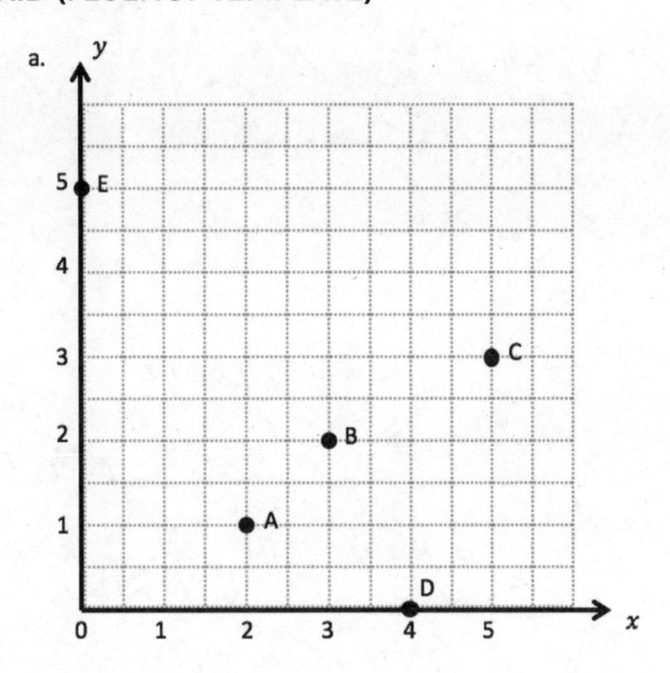

b.

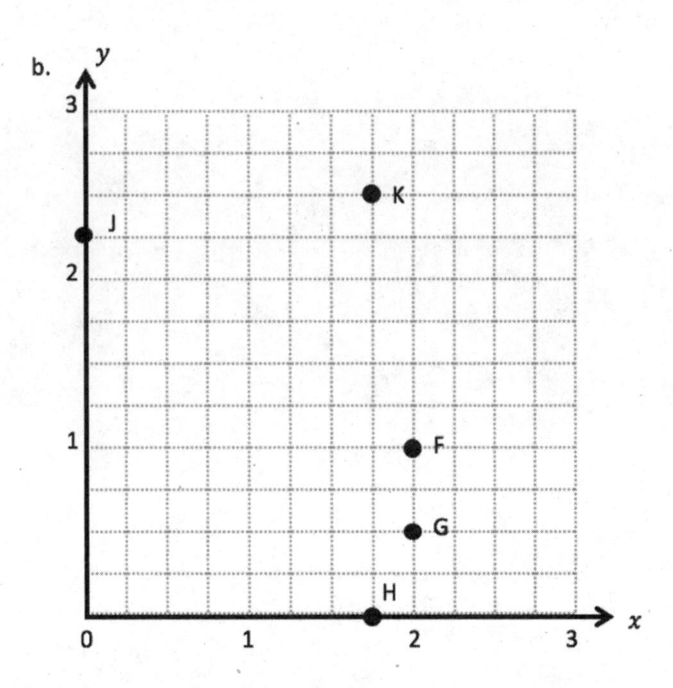

PROBLEM SET

Battleship Rules

Goal: To sink all of your opponent's ships by correctly guessing their coordinates.

Materials

- 1 grid sheet (per person/per game)
- Red crayon/marker for hits
- Black crayon/marker for misses
- Folder to place between players

Ships

- Each player must mark 5 ships on the grid.
  Aircraft carrier—plot 5 points.
  Battleship—plot 4 points.
  Cruiser—plot 3 points.
  Submarine—plot 3 points.
  Patrol boat—plot 2 points.

Setup

- With your opponent, choose a unit length and fractional unit for the coordinate plane.
- Label the chosen units on both grid sheets.
- Secretly select locations for each of the 5 ships on your My Ships grid.
  All ships must be placed horizontally or vertically on the coordinate plane.
  Ships can touch each other, but they may not occupy the same coordinate.

Play

- Players take turns firing one shot to attack enemy ships.
- On your turn, call out the coordinates of your attacking shot. Record the coordinates of each attack shot.
- Your opponent checks his/her My Ships grid. If that coordinate is unoccupied, your opponent says, "Miss." If you named a coordinate occupied by a ship, your opponent says, "Hit."
- Mark each attempted shot on your Enemy Ships grid. Mark a black **X** on the coordinate if your opponent says, "Miss." Mark a red ✓ on the coordinate if your opponent says, "Hit."

- On your opponent's turn, if he/she hits one of your ships, mark a red ✓ on that coordinate of your My Ships grid. When one of your ships has every coordinate marked with a ✓, say, "You've sunk my [name of ship]."

Victory

- The first player to sink all (or the most) opposing ships, wins.

My Ships

- Draw a red ✓ over any coordinate your opponent hits.
- Once all of the coordinates of any ship have been hit, say, "You've sunk my [name of ship]."

Aircraft carrier—5 points
Battleship—4 points
Cruiser—3 points
Submarine—3 points
Patrol boat—2 points

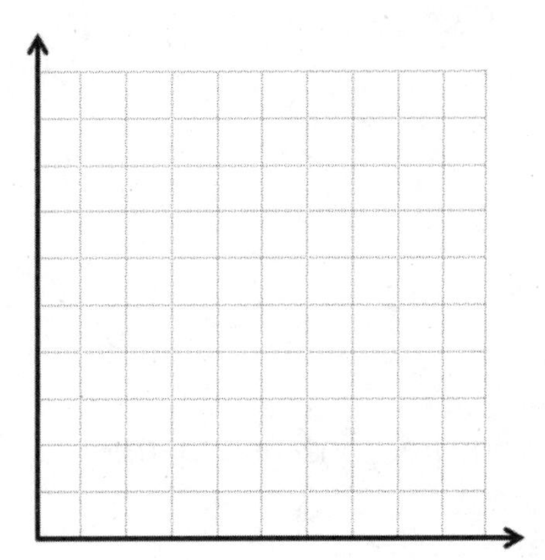

Attack Shots

- Record the coordinates of each shot below and whether it was a ✓ (hit) or an **X** (miss).

(______, ______) (______, ______)
(______, ______) (______, ______)
(______, ______) (______, ______)
(______, ______) (______, ______)
(______, ______) (______, ______)
(______, ______) (______, ______)
(______, ______) (______, ______)
(______, ______) (______, ______)

Enemy Ships

- Draw a black **X** on the coordinate if your opponent says, "Miss."
- Draw a red ✓ on the coordinate if your opponent says, "Hit."
- Draw a circle around the coordinates of a sunken ship.

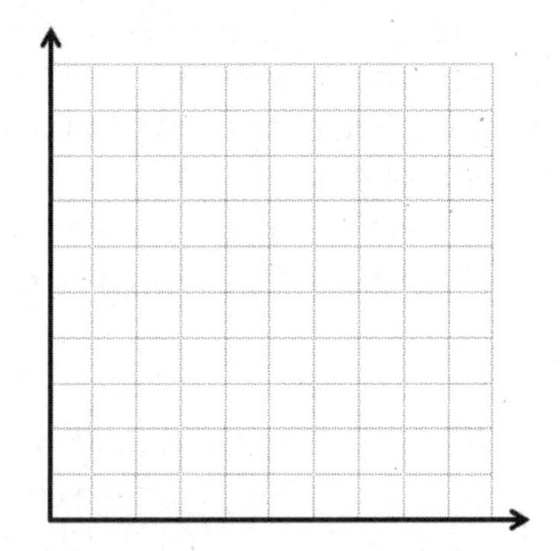

Lesson 5

Word Problem

A company has developed a new game. Cartons are needed to ship 40 games at a time. Each game is 2 inches high by 7 inches wide by 14 inches long.

How would you recommend packing the board games in the carton? What are the dimensions of a carton that could ship 40 board games with no extra room in the box?

Name: ______________________________ **Date:** ____________

GRADE 5 / MISSION 6 / LESSON 5

Exit Ticket

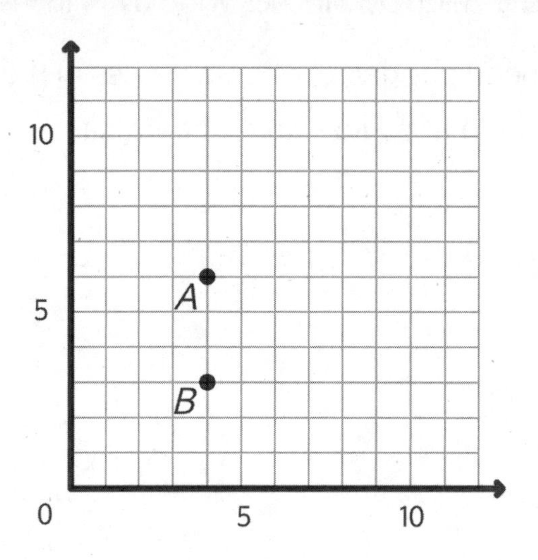

1. Use a straight edge to construct a line that goes through points *A* and *B*. Label the line *l*.

2. Which axis is parallel to line *l* ? ________________

   Which axis is perpendicular to line *l* ? ________________

3. Plot two more points on line *l*. Name them *C* and *D*.

4. Give the coordinates of each point below.

   A : __________ *B* : __________

   C : __________ *D* : __________

5. Give the coordinates of another point that falls on line *l* with a *y*-coordinate greater than 20. __________

COORDINATE PLANE PRACTICE (CONCEPT EXPLORATION TEMPLATE)

Point	x	y	(x, y)
H			
I			
J			
K			
L			

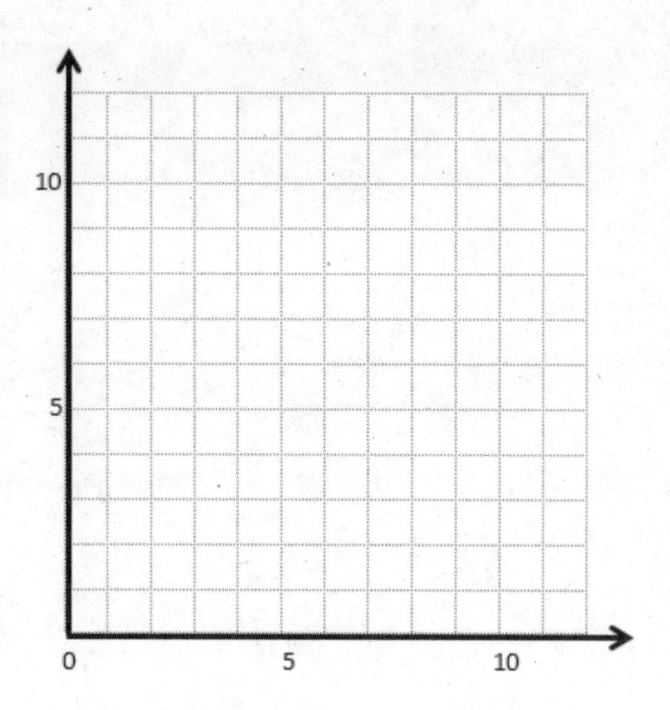

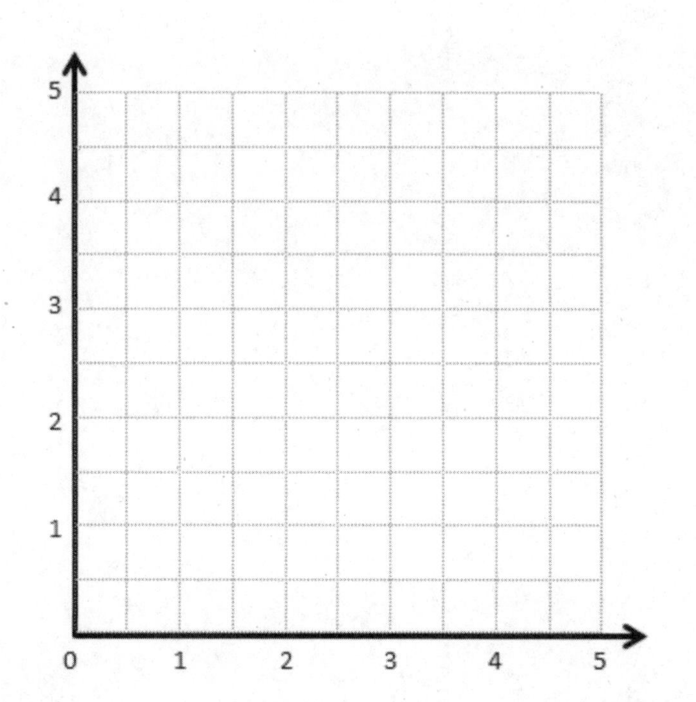

Point	x	y	(x, y)
D	$2\frac{1}{2}$	0	$\left(2\frac{1}{2}, 0\right)$
E	$2\frac{1}{2}$	2	$\left(2\frac{1}{2}, 2\right)$
F	$2\frac{1}{2}$	4	$\left(2\frac{1}{2}, 4\right)$

Lesson 6

Word Problem

Adam built a toy box for his children's wooden blocks.

a. If the inside dimensions of the box are 18 inches by 12 inches by 6 inches, what is the maximum number of 2-inch wooden cubes that will fit in the toy box?

b. What if Adam had built the box 16 inches by 9 inches by 9 inches? What is the maximum number of 2-inch wooden cubes that would fit in this size box?

Name: ______________________ **Date:** __________

GRADE 5 / MISSION 6 / LESSON 6

Exit Ticket

You'll need two colored pencils for this exit ticket.

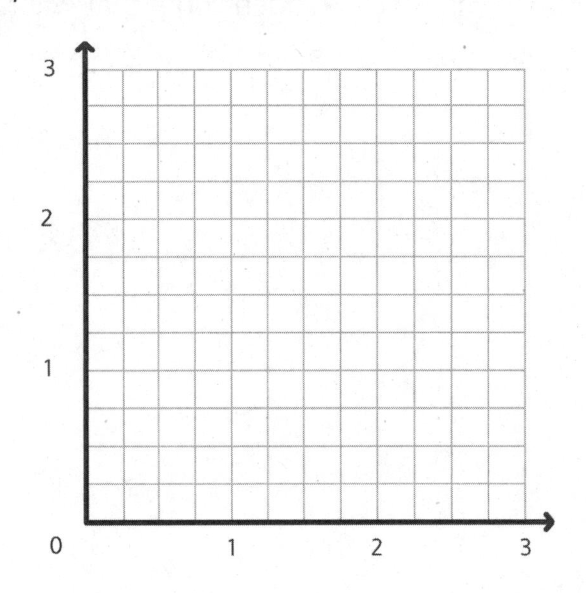

1. Plot the point $H\left(2\frac{1}{2}, 1\frac{1}{2}\right)$.

2. Line l passes through point H and is parallel to the y-axis. Construct line l.

3. Construct line m such that the y-coordinate of every point is $\frac{3}{4}$.

4. Line m is ______ units from the x-axis.

5. Give the coordinates of the point on line m that is $\frac{1}{2}$ unit from the y-axis.

6. With one colored pencil, shade the portion of the plane that is less than $\frac{3}{4}$ unit from the x-axis.

7. With another colored pencil, shade the portion of the plane that is less than $2\frac{1}{2}$ units from the y-axis.

8. Plot a point that lies in the double shaded region. Give the coordinates of the point.

MILLIONS THROUGH THOUSANDTHS PLACE VALUE CHART (FLUENCY TEMPLATE)

1,000,000	100,000	10,000	1,000	100	10	1	•	$\frac{1}{10}$	$\frac{1}{100}$	$\frac{1}{1000}$
Millions	Hundred Thousands	Ten Thousands	Thousands	Hundreds	Tens	Ones	•	Tenths	Hundredths	Thousandths
							•			
							•			
							•			
							•			
							•			
							•			
							•			
							•			
							•			

COORDINATE PLANE (CONCEPT EXPLORATION TEMPLATE)

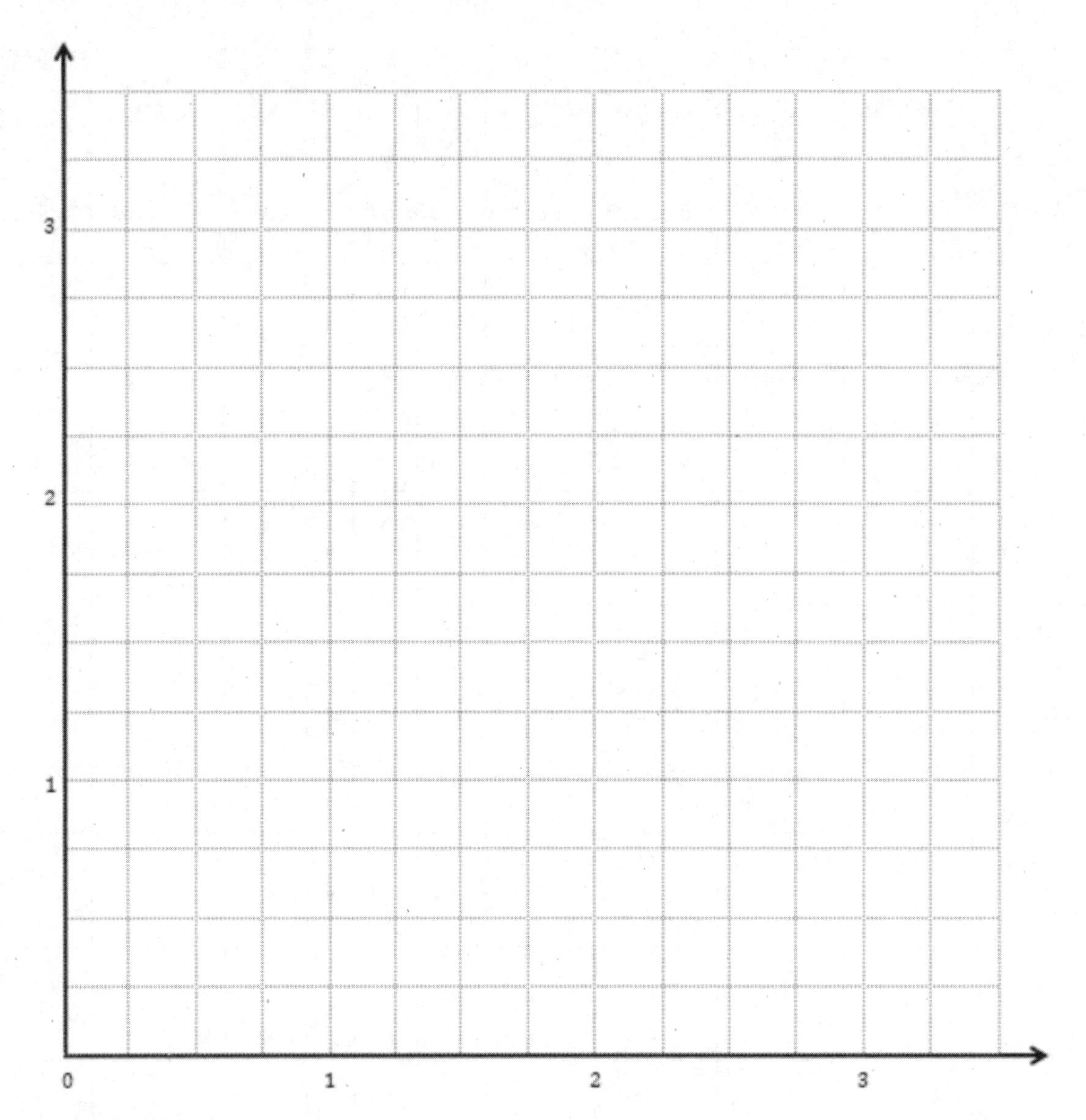

Point	x	y	(x, y)
A			
B			
C			

Point	x	y	(x, y)
D			
E			
C			

Lesson 7

Word Problem

An orchard charges $0.85 to ship a quarter kilogram of grapefruit. Each grapefruit weighs approximately 165 grams. How much will it cost to ship 40 grapefruits?

Name: ______________________ Date: __________

GRADE 5 / MISSION 6 / LESSON 7

Exit Ticket

Complete the chart. Then, plot the points on the coordinate plane.

x	y	(x, y)
0	4	
2	6	
3	7	
7	11	

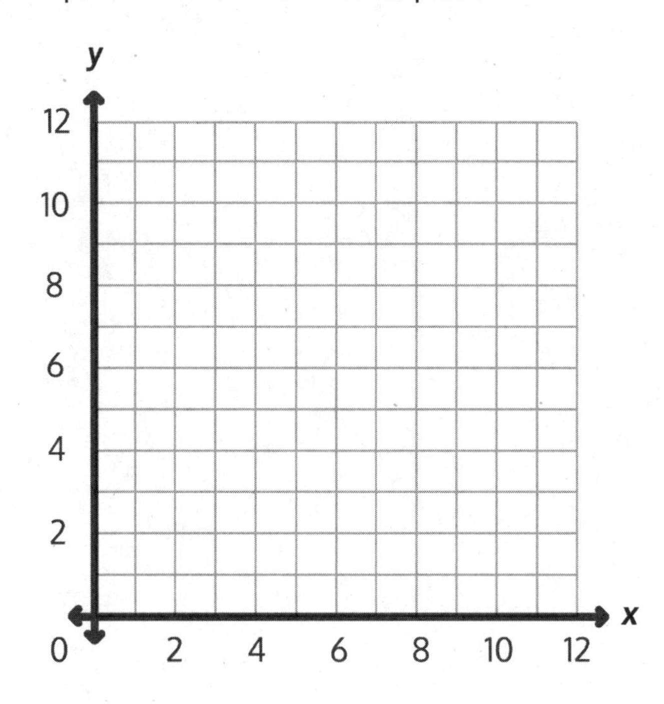

1. Use a straight edge to draw a line connecting these points.
2. Write a rule to show the relationship between the *x*-and *y*-coordinates for points on the line.

   ____

   ____

3. Name two other points that are also on this line.

   __________ __________

COORDINATE GRID (FLUENCY TEMPLATE)

a.

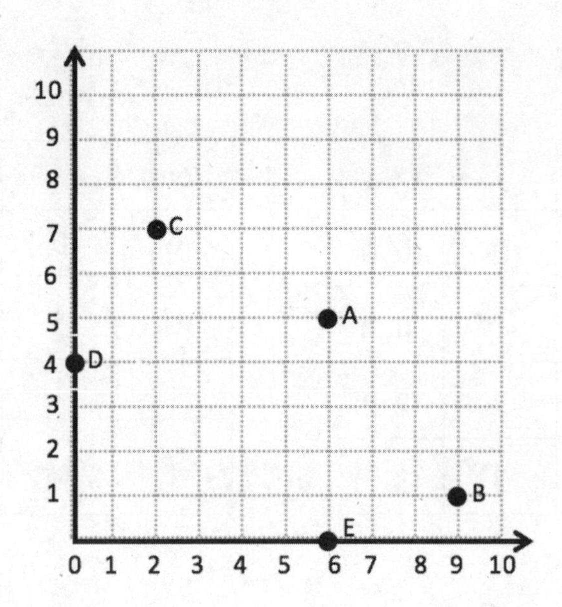

b.

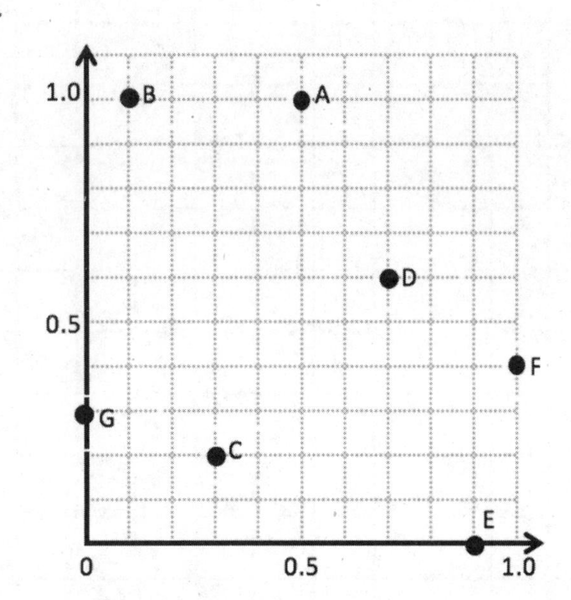

COORDINATE PLANE (CONCEPT EXPLORATION TEMPLATE) (PAGE 1 OF 2)

1.

a.

Point	x	y	(x, y)
A	0	0	(0, 0)
B	1	1	(1, 1)
C	2	2	(2, 2)
D	3	3	(3, 3)

b.

Point	x	y	(x, y)
G	0	3	(0, 3)
H	$\frac{1}{2}$	$3\frac{1}{2}$	$\left(\frac{1}{2}, 3\frac{1}{2}\right)$
I	1	4	(1, 4)
J	$1\frac{1}{2}$	$4\frac{1}{2}$	$\left(1\frac{1}{2}, 4\frac{1}{2}\right)$

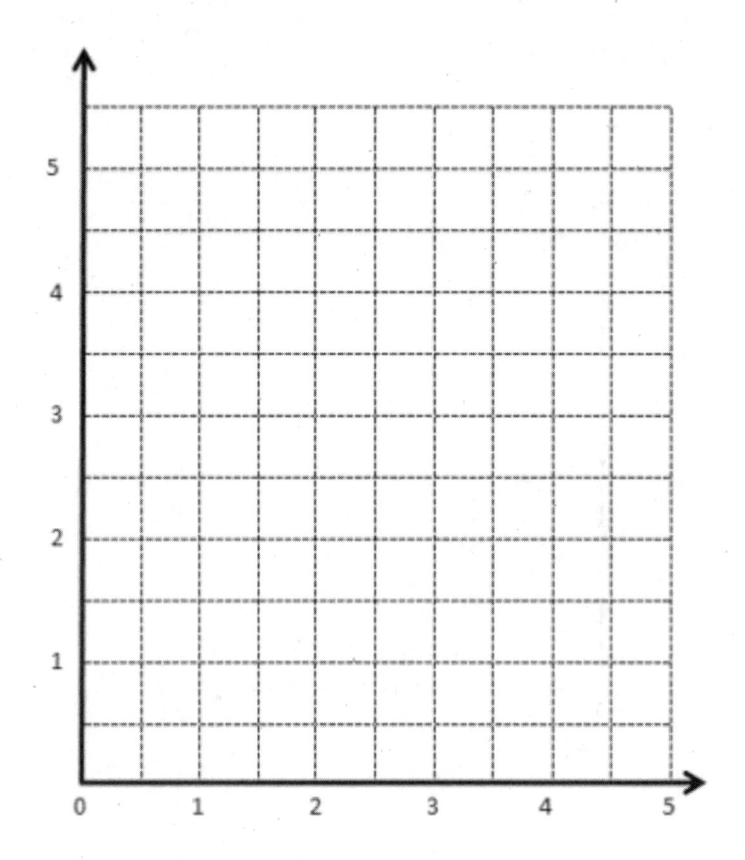

COORDINATE PLANE (CONCEPT EXPLORATION TEMPLATE) (PAGE 2 OF 2)

2.

a.

Point	(x, y)
L	(0, 3)
M	(2, 3)
N	(4, 3)

b.

Point	(x, y)
O	(0, 0)
P	(1, 2)
Q	(2, 4)

c.

Point	(x, y)
R	$\left(1, \frac{1}{2}\right)$
S	$\left(2, 1\frac{1}{2}\right)$
T	$\left(3, 2\frac{1}{2}\right)$

d.

Point	(x, y)
U	(1, 3)
V	(2, 6)
W	(3, 9)

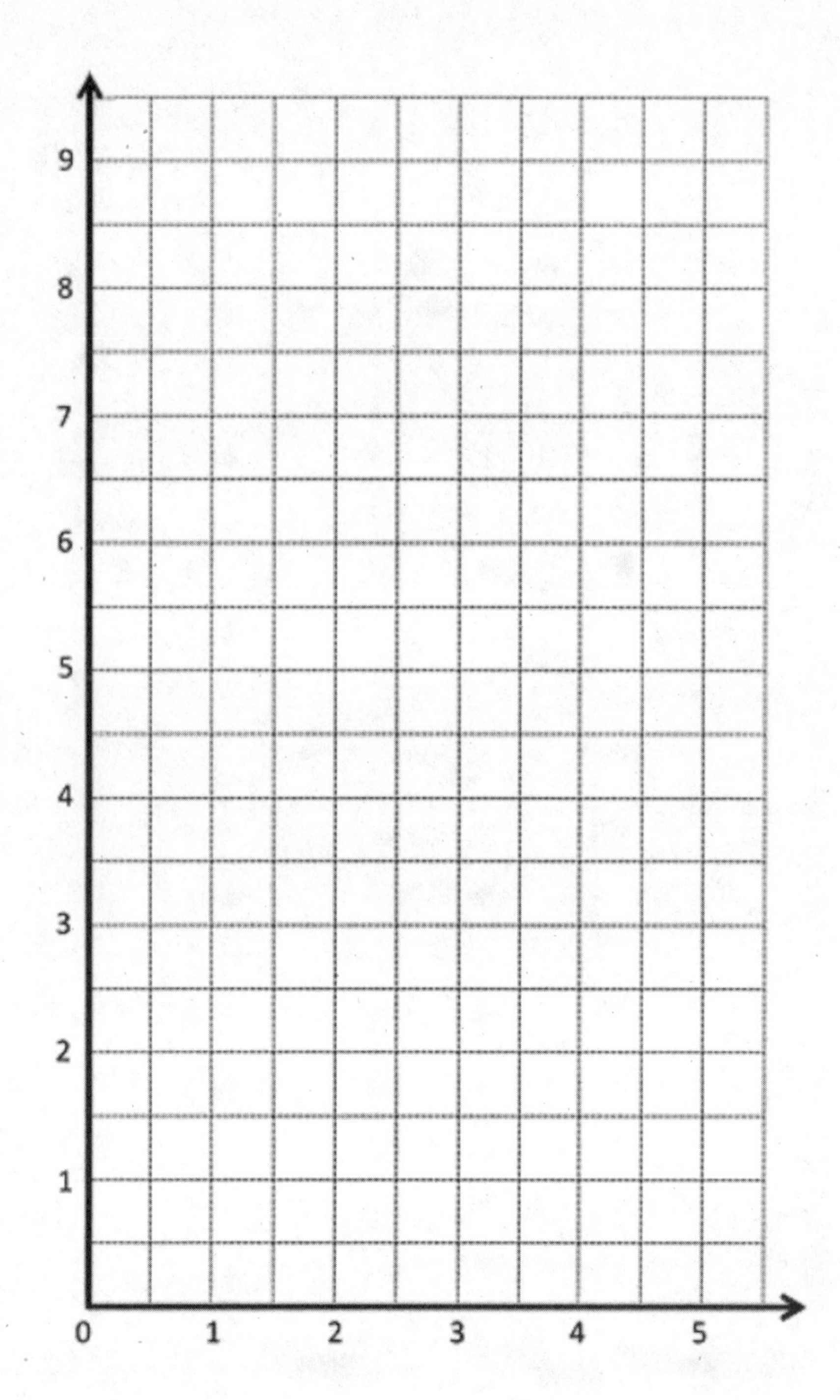

Lesson 8

Word Problem

The coordinate pairs listed locate points on two different lines. Write a rule that describes the relationship between the *x*- and *y*-coordinates for each line.

Line l: $\left(3\frac{1}{2}, 7\right)$, $\left(1\frac{2}{3}, 3\frac{1}{3}\right)$, $(5, 10)$

Line m: $\left(\frac{6}{3}, 1\right)$, $\left(3\frac{1}{2}, 1\frac{3}{4}\right)$, $\left(13, 6\frac{1}{2}\right)$

Name: ______________________ Date: __________

GRADE 5 / MISSION 6 / LESSON 8

Exit Ticket

1. Complete this table with values for *y* such that each *y*-coordinate is 5 more than 2 times as much as its corresponding *x*-coordinate.

   a. Plot each point on the coordinate plane.

   b. Use a straight edge to draw a line connecting these points.

   c. Name 2 other points that fall on this line with *y*-coordinates greater than 25.

x	y	(x, y)
0		
2		
3.5		

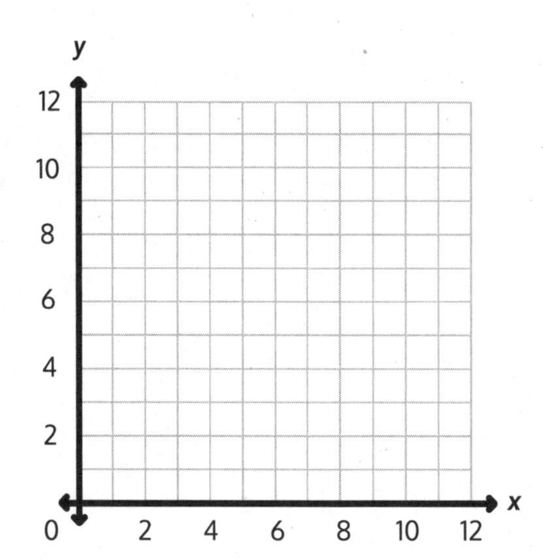

COORDINATE GRID INSERT (FLUENCY TEMPLATE)

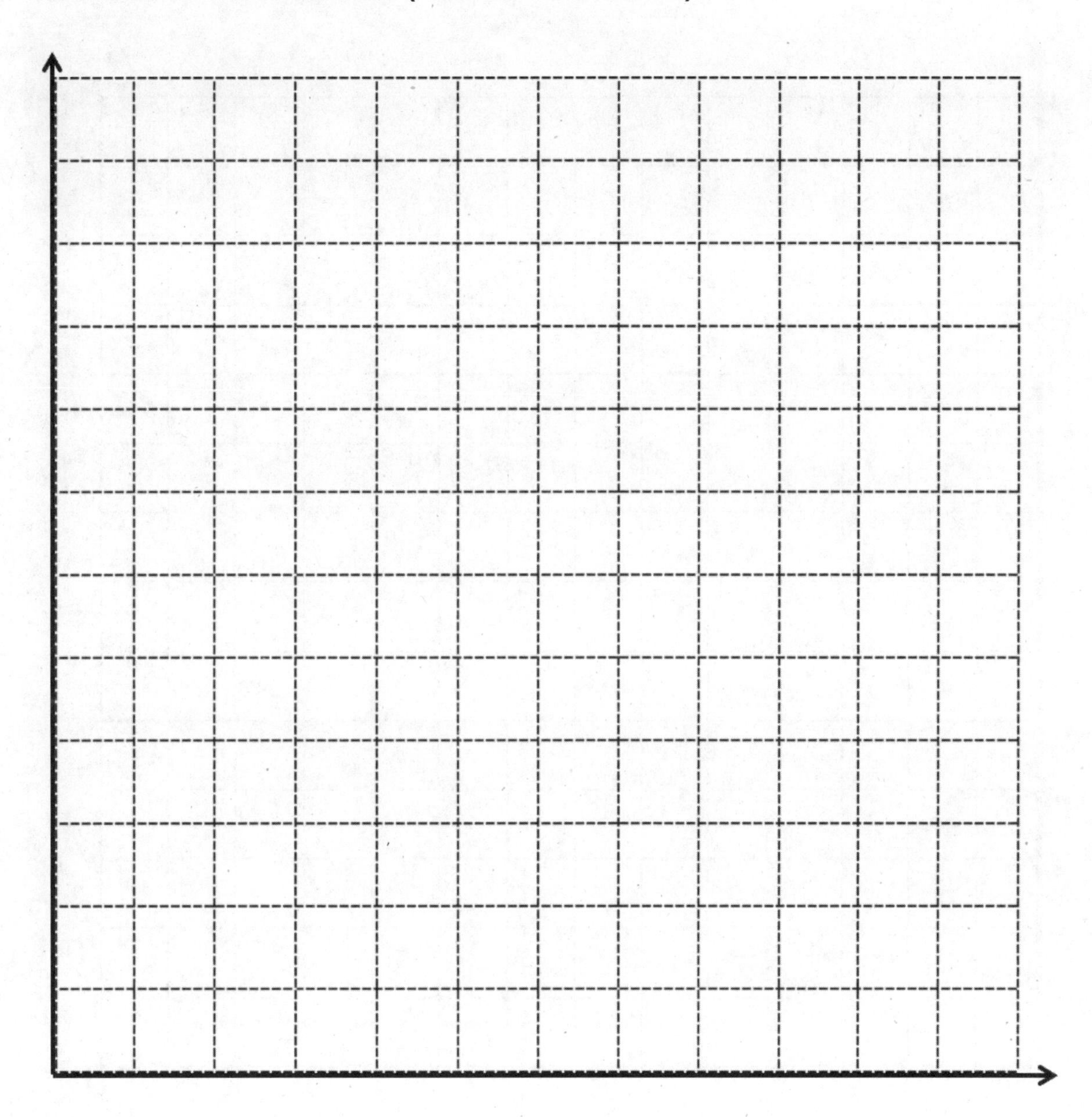

COORDINATE PLANE (CONCEPT EXPLORATION TEMPLATE)

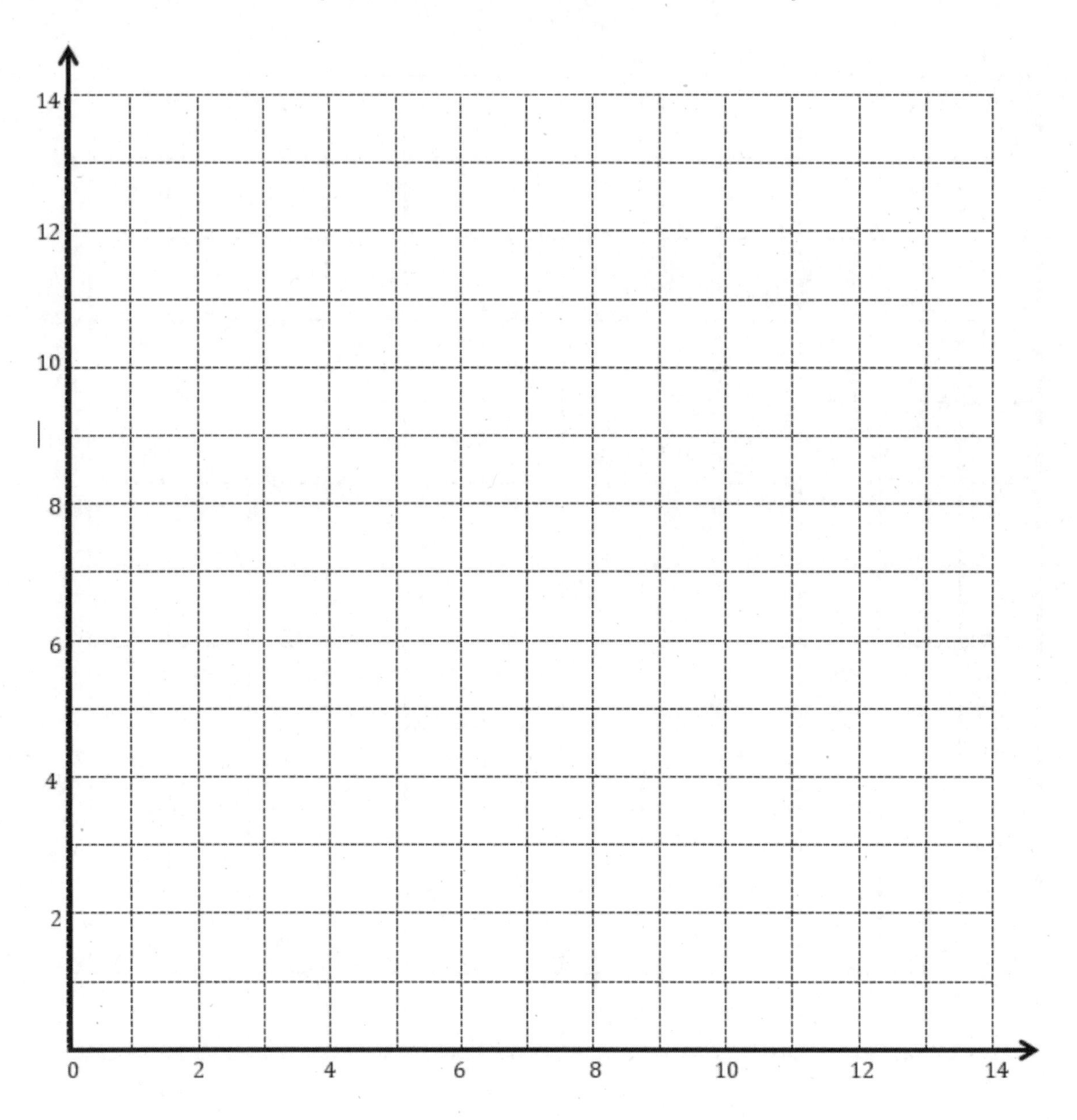

Line a:		
x	y	(x, y)

Line b:		
x	y	(x, y)

Line c:		
x	y	(x, y)

Lesson 9

Word Problem

Maggie spent $46.20 to buy pencil sharpeners for her gift shop. If each pencil sharpener costs 60 cents, how many pencil sharpeners did she buy? Solve by using the standard algorithm.

Name: ______________________ Date: ____________

GRADE 5 / MISSION 6 / LESSON 9

Exit Ticket

1. Complete the table for the given rules.

   Then, construct lines *l* and *m* on the coordinate plane.

Line *l*

Rule: y *is* 5 *more than* x

x	y	(x, y)
0		
1		
2		
4		

Line *m*

Rule: y *is* 5 *times as much as* x

x	y	(x, y)
0		
1		
2		
4		

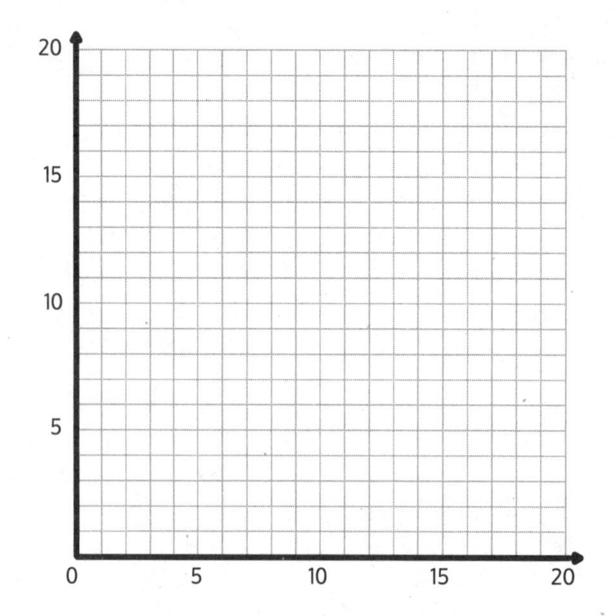

COORDINATE PLANE (CONCEPT EXPLORATION TEMPLATE) (PAGE 1 OF 2)

Line l

Rule: y is 2 more than x

x	y	(x, y)
1		
5		
10		
15		

Line m

Rule: y is 5 more than x

x	y	(x, y)
0		
5		
10		
15		

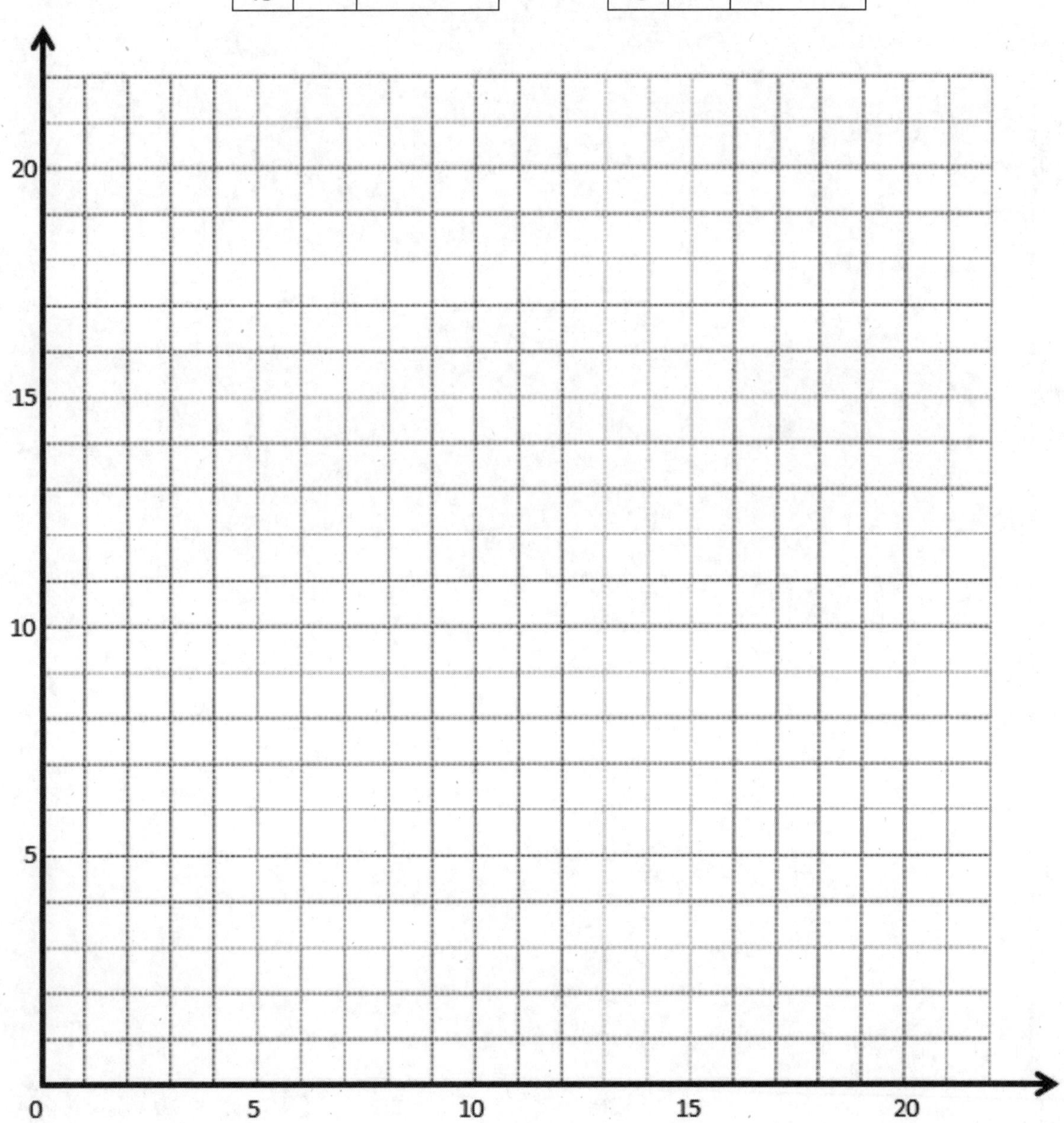

COORDINATE PLANE (CONCEPT EXPLORATION TEMPLATE) (PAGE 2 OF 2)

Line *p*

Rule: y is x times 2

x	y	(x, y)

Line *q*

Rule: y is x times 3

x	y	(x, y)

20

15

10

5

0 5 10 15 20

Lesson 10

Word Problem

A 12-man relay team runs a 45 km race. Each member of the team runs an equal distance. How many kilometers does each team member run? One lap around the track is 0.75 km. How many laps does each team member run during the race?

Name: ______________________ Date: ____________

GRADE 5 / MISSION 6 / LESSON 10

Exit Ticket

1. Use the coordinate plane below to complete the following tasks.

   Line p represents the rule *x and y are equal*.

   a. Construct a line, a, that is parallel to line p and contains point A.

   b. Name 3 points on line a.

   c. Identify a rule to describe line a.

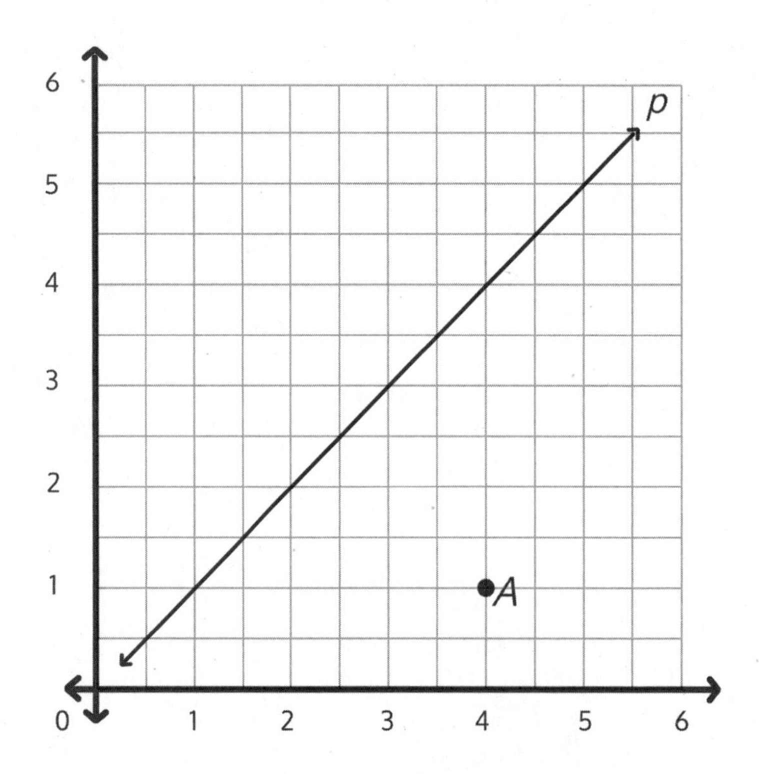

COORDINATE PLANE (CONCEPT EXPLORATION TEMPLATE) (PAGE 1 OF 2)

Line *p*

Rule: y is 0 more than x

x	y	(x, y)
0		
5		
10		
15		

Line *b*

Rule: ______________

x	y	(x, y)
7		
10		
13		
18		

Line *c*

Rule: ______________

x	y	(x, y)
2		
4		
8		
11		

Line *d*

Rule: ______________

x	y	(x, y)
5		
7		
12		
15		

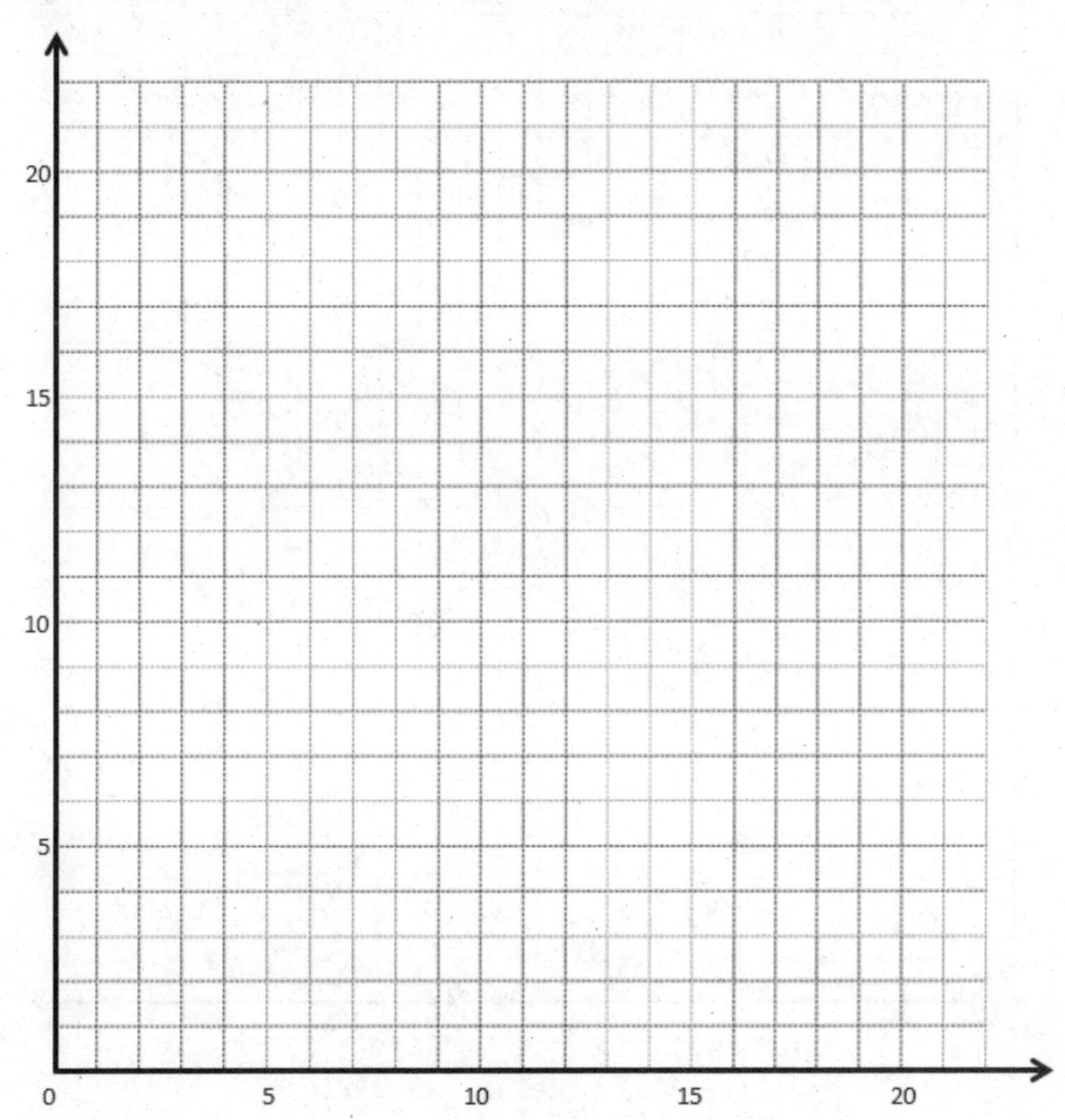

COORDINATE PLANE (CONCEPT EXPLORATION TEMPLATE) (PAGE 2 OF 2)

Line g Rule: ____________________

x	y	(x, y)
1		
2		
5		
7		

Line h Rule: ____________________

x	y	(x, y)
3		
6		
12		
15		

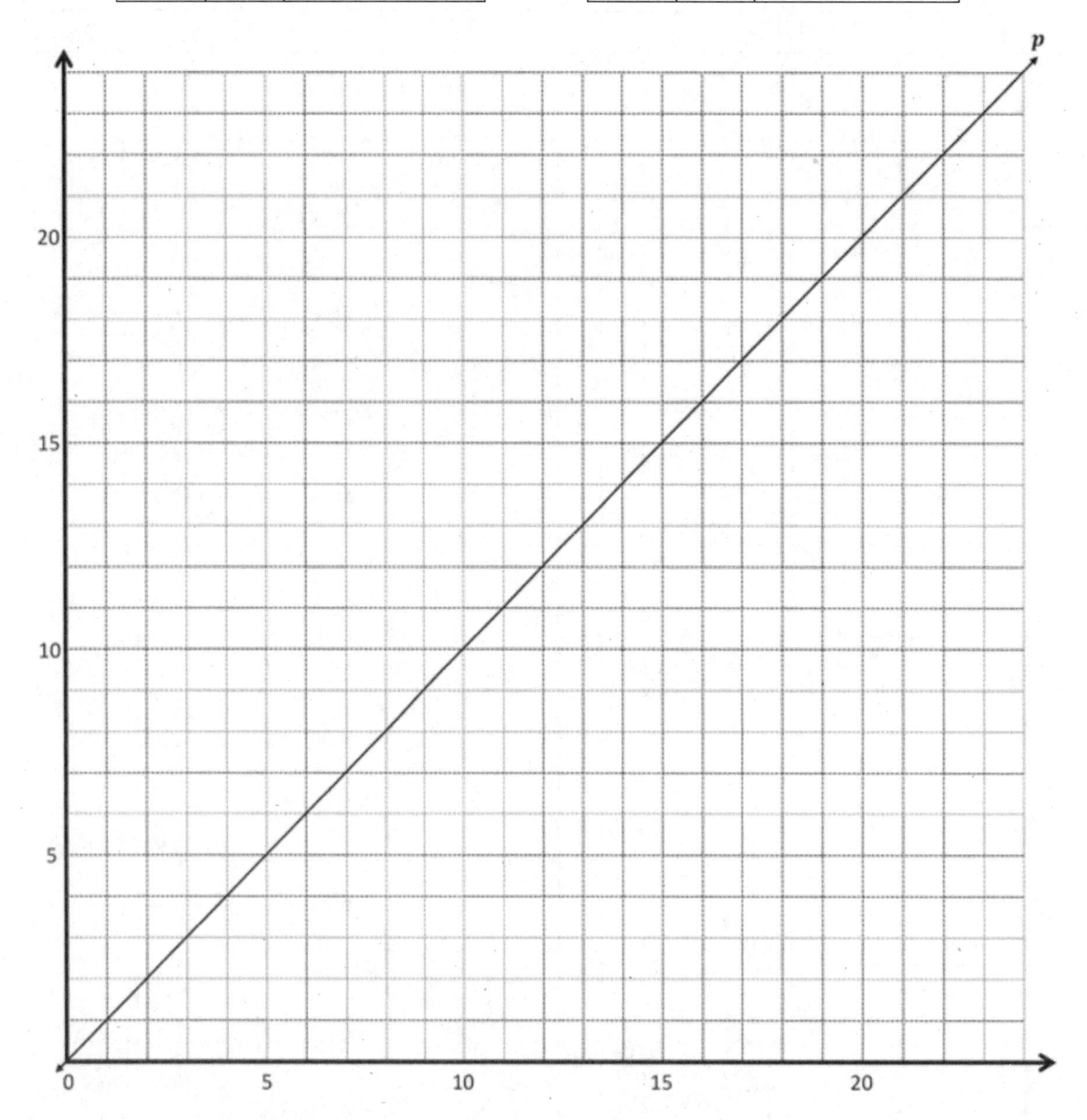

Lesson 11

Word Problem

Michelle has 3 kg of strawberries that she divided equally into small bags with $\frac{1}{5}$ kg in each bag.

a. How many bags of strawberries did she make?

b. She gave a bag to her friend, Sarah. Sarah ate half of her strawberries. How many grams of strawberries does Sarah have left?

Name: ______________________ Date: __________

GRADE 5 / MISSION 6 / LESSON 11

Exit Ticket

1. Complete the tables for the given rules.

Line *l*

Rule: *Triple x*

x	y	(x, y)
0		
1		
2		
3		

Line *m*

Rule: *Triple x, and then add 1*

x	y	(x, y)
0		
1		
2		
3		

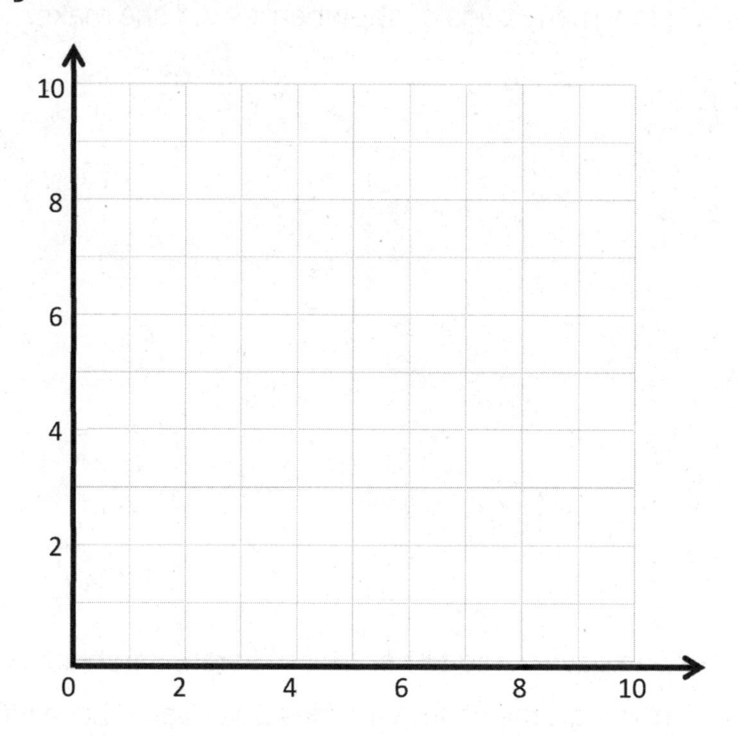

a. Draw each line on the coordinate plane above.

b. Compare and contrast these lines.

2. Circle the point(s) that the line for the rule *multiply x by $\frac{1}{3}$, and then add 1* would contain.

$\left(0, \frac{1}{2}\right)$ $\left(1, 1\frac{1}{3}\right)$ $\left(2, 1\frac{2}{3}\right)$ $\left(3, 2\frac{1}{2}\right)$

COORDINATE PLANE (CONCEPT EXPLORATION TEMPLATE)

Line *l*

Rule: *Triple x*

x	y	(x, y)
0		
1		
2		
4		

Line *m*

Rule: *Triple x, and then add 3*

x	y	(x, y)
0		
1		
2		
3		

Line *n*

Rule: *Triple x, and then subtract 2*

x	y	(x, y)
1		
2		
3		
4		

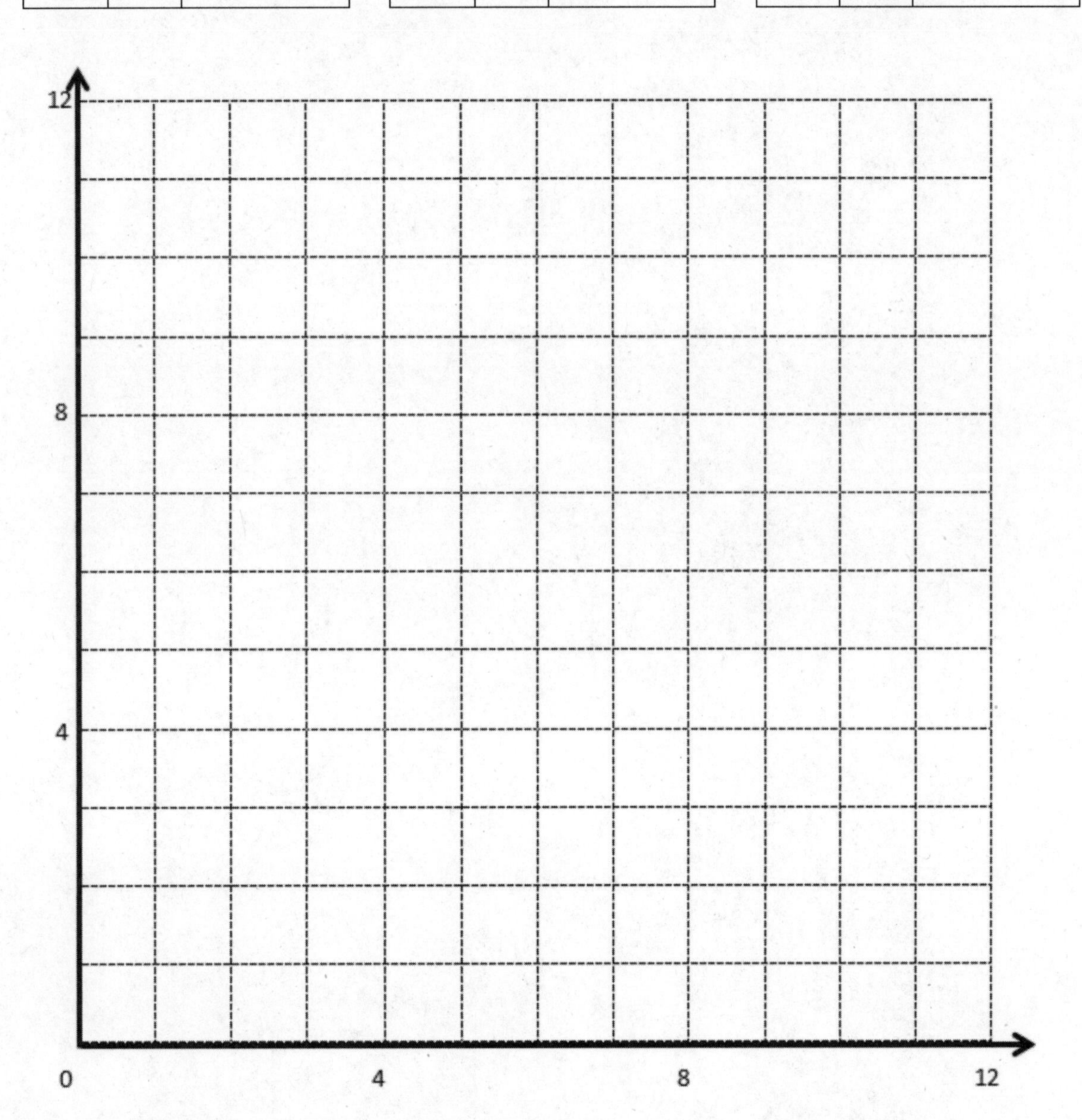

Lesson 12

Word Problem

Mr. Jones had 640 books. He sold $\frac{1}{4}$ of them for $2.00 each in the month of September. He sold half of the remaining books in October. Each book he sold in October earned $\frac{3}{4}$ of what each book sold for in September. How much money did Mr. Jones earn selling books? Show your thinking with a tape diagram.

Name: ______________________ **Date:** __________

GRADE 5 / MISSION 6 / LESSON 12

Exit Ticket

Write the rule for the line that contains the points $\left(0, 1\frac{1}{2}\right)$ and $\left(1\frac{1}{2}, 3\right)$.

a. Identify 2 more points on this line. Draw the line on the grid.

Point	x	y	(x, y)
B			
C			

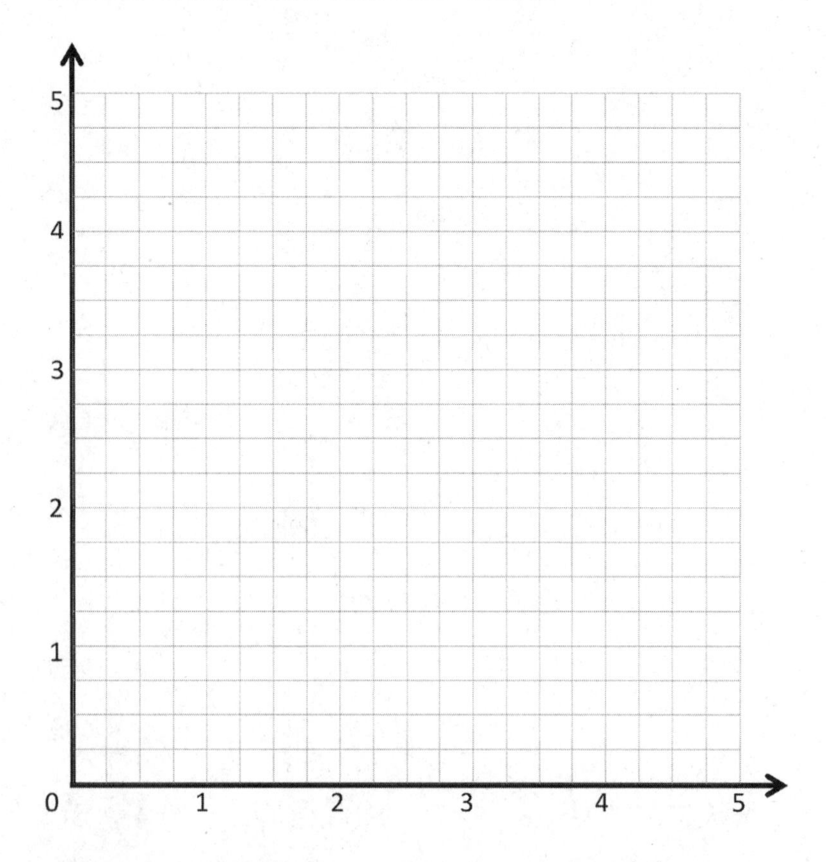

b. Write a rule for a line that is parallel to $\overleftrightarrow{BC}$ and goes through point $\left(1, \frac{1}{2}\right)$.

COORDINATE PLANE (CONCEPT EXPLORATION TEMPLATE)

Line *l*

Rule: ____________________

Point	x	y	(x, y)
A	$1\frac{1}{2}$	3	$\left(1\frac{1}{2}, 3\right)$
B			
C			
D			

Line *m*

Rule: ____________________

Point	x	y	(x, y)
A			
E			
F			
G			

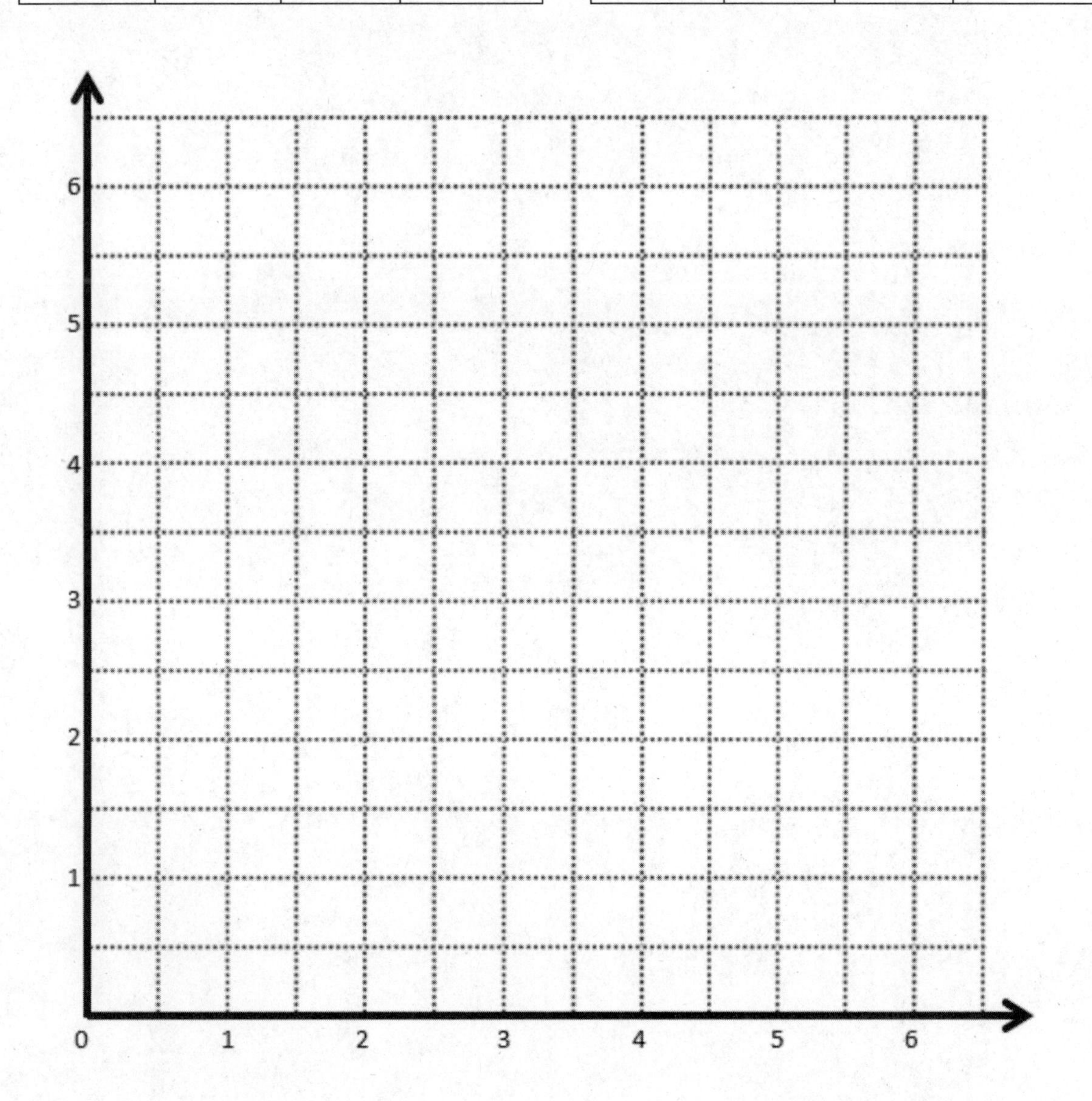

Lesson 13

Name: ____________________ **Date:** __________

GRADE 5 / MISSION 6 / LESSON 13

Exit Ticket

Use your straightedge to draw a segment parallel to each segment through the given point.

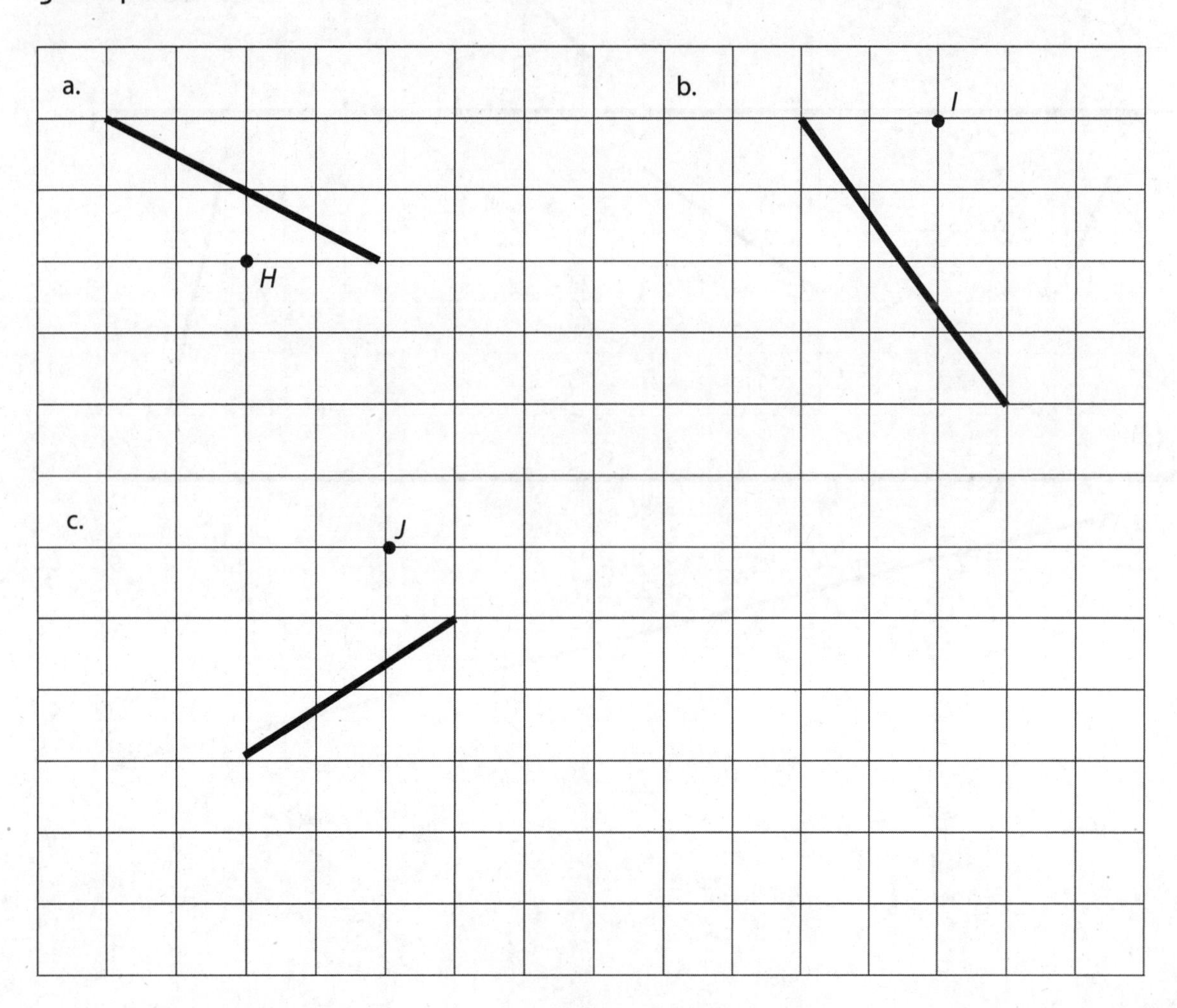

RECORDING SHEET (CONCEPT EXPLORATION TEMPLATE 2)

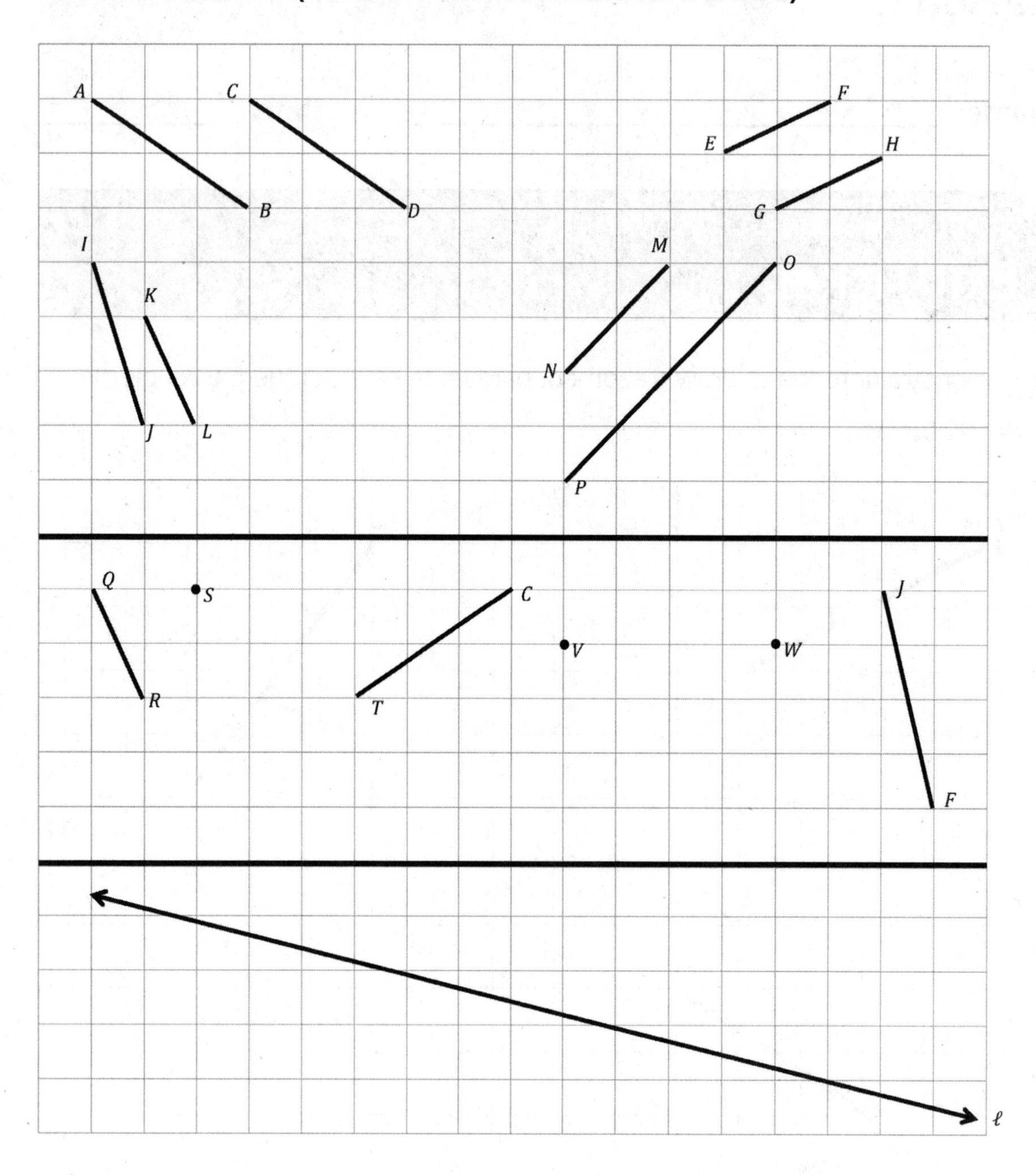

Lesson 14

Word Problem

Drew's fish tank measures 32 cm by 22 cm by 26 cm. He pours 20 liters of water into it, and some water overflows the tank. Find the volume of water, in milliliters, that overflows.

Name: ______________________ Date: ____________

GRADE 5 / MISSION 6 / LESSON 14

Exit Ticket

1. Use the coordinate plane below to complete the following tasks.

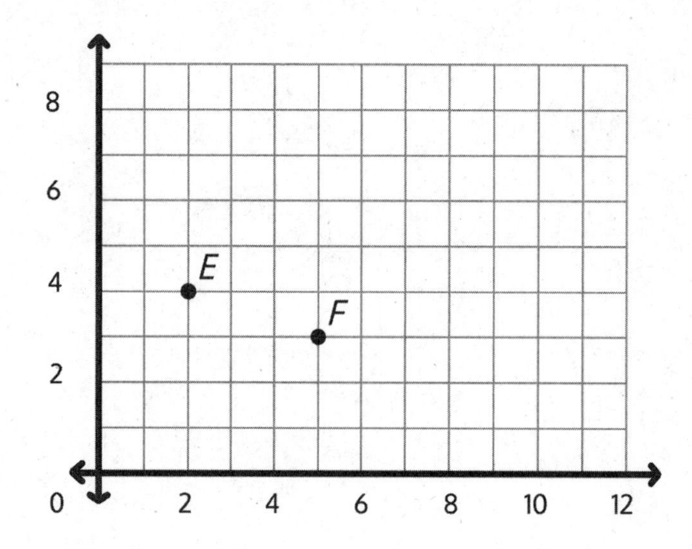

a. Identify the locations of E and F.

E: (_____, _____) F: (_____, _____)

b. Draw $\overleftrightarrow{EF}$.

c. Generate coordinate pairs for L and M, such that $\overleftrightarrow{EF} \parallel \overleftrightarrow{LM}$.

L: (_____, _____) M: (_____, _____)

d. Draw $\overleftrightarrow{LM}$.

COORDINATE PLANE (CONCEPT EXPLORATION TEMPLATE)

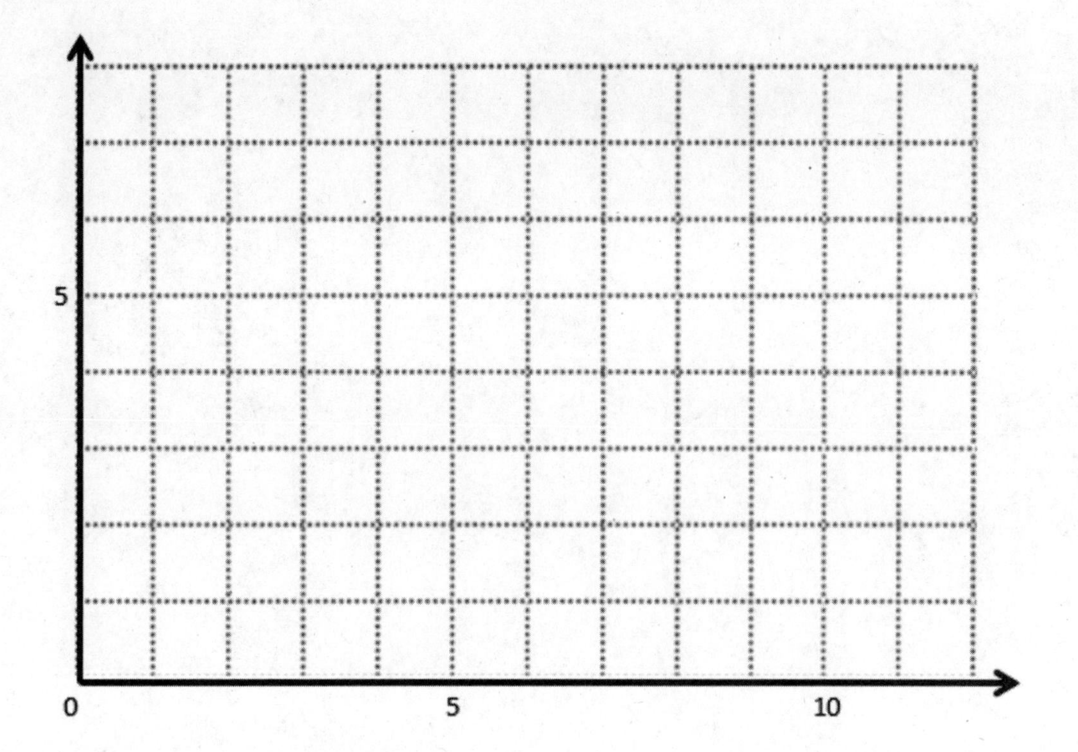

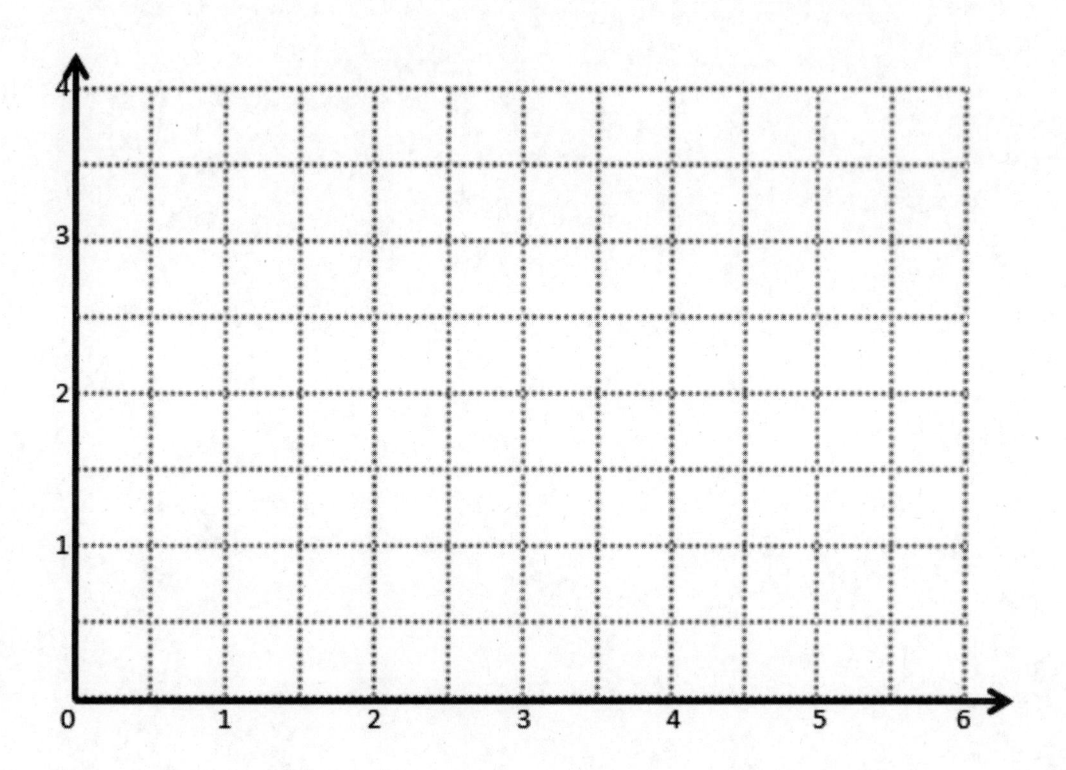

Lesson 15

Name: ______________________ Date: __________

GRADE 5 / MISSION 6 / LESSON 15

Exit Ticket

1. Draw a segment perpendicular to each given segment. Show your thinking by sketching triangles as needed.

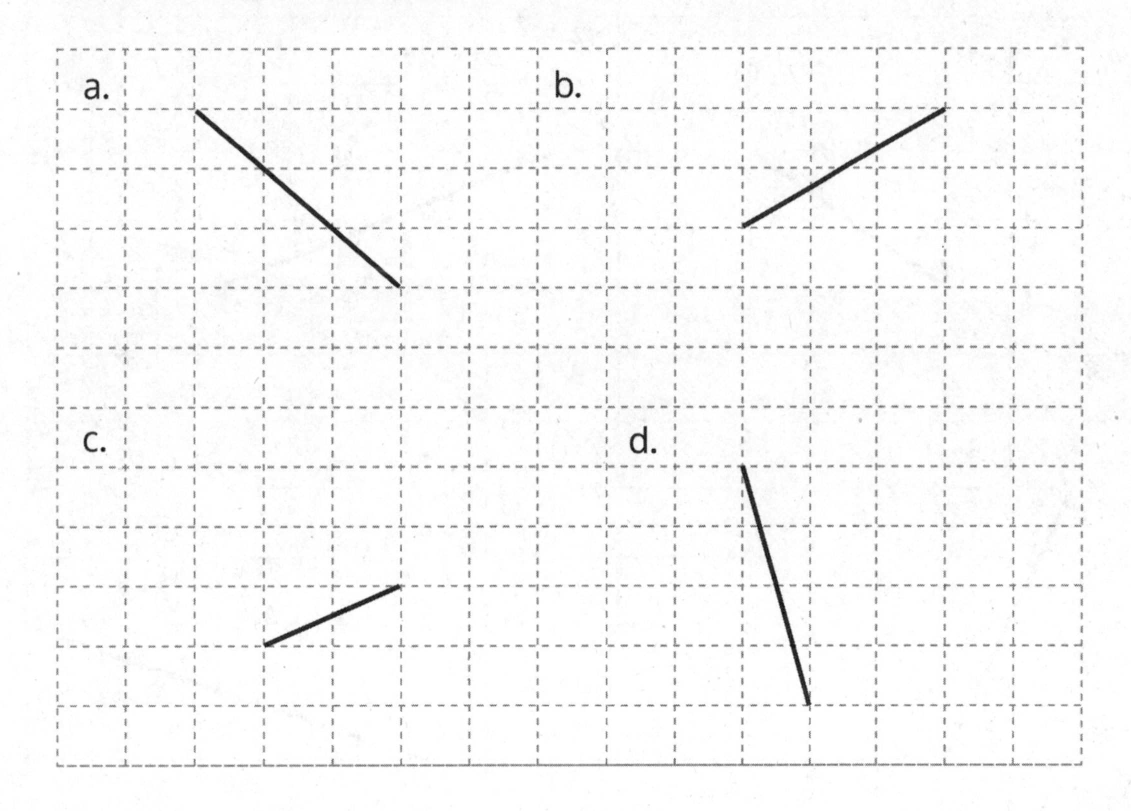

RECORDING SHEET (CONCEPT EXPLORATION TEMPLATE 1)

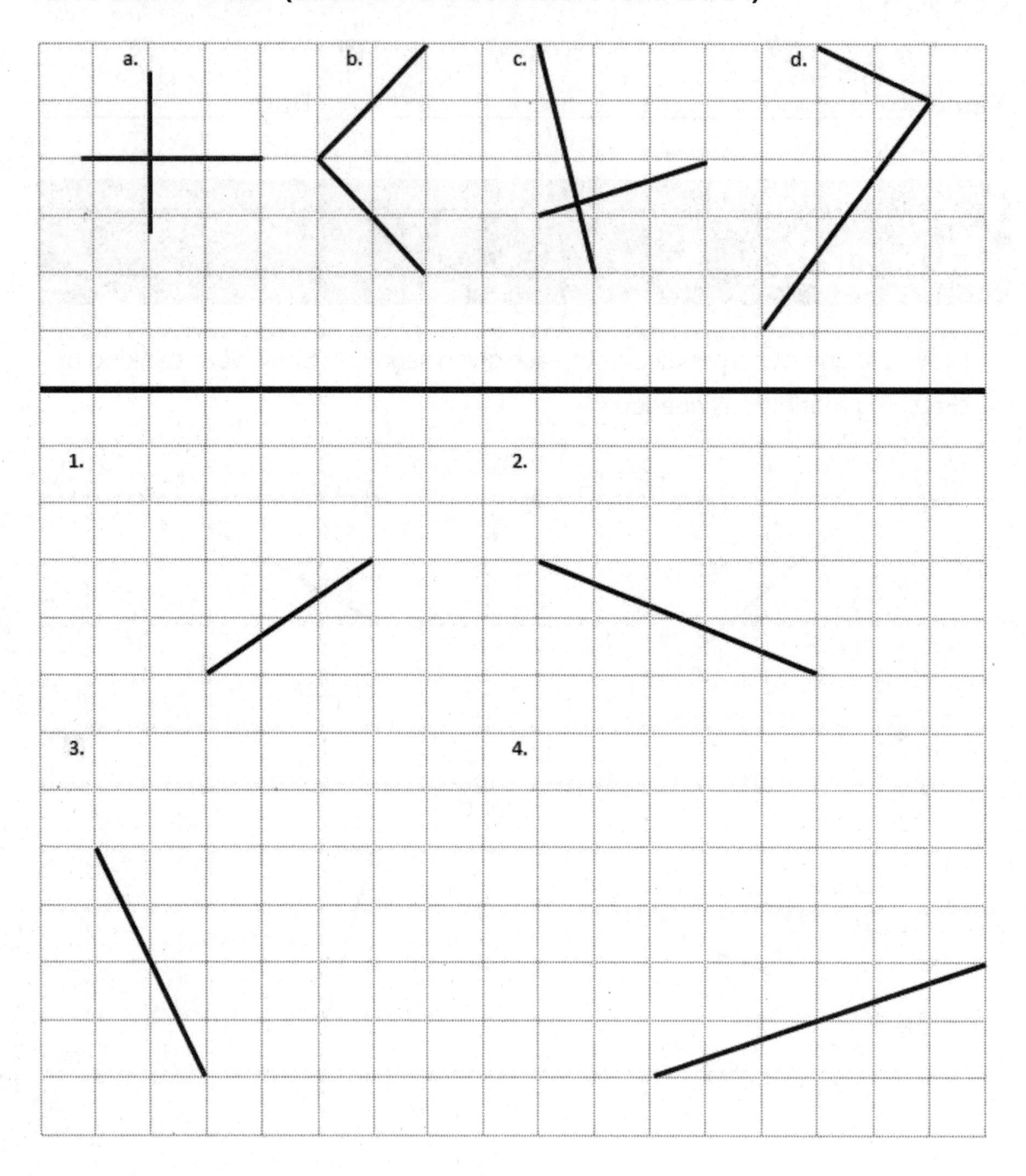

Lesson 16

Word Problem

a. Complete the table for the rule *y is 1 more than half x*, graph the coordinate pairs, and draw a line to connect them.

b. Give the y-coordinate for the point on this line whose x-coordinate is $42\frac{1}{4}$.

Extension: Give the x-coordinate for the point on this line whose y-coordinate is $5\frac{1}{2}$.

x	y
$\frac{1}{2}$	
$1\frac{1}{2}$	
$2\frac{1}{4}$	
3	

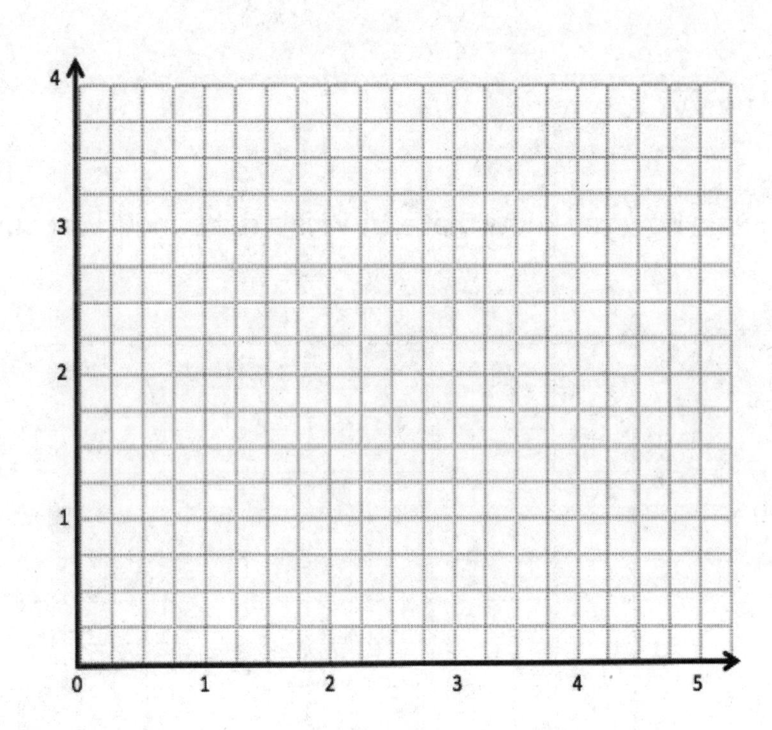

Name: ______________________________ Date: ____________

GRADE 5 / MISSION 6 / LESSON 16

Exit Ticket

Use the coordinate plane below to complete the following tasks.

a. Draw $\overline{UV}$.

b. Plot point $W\left(4\frac{1}{2}, 6\right)$.

c. Draw $\overline{VW}$.

d. Explain how you know that $\angle UVW$ is a right angle without measuring it.

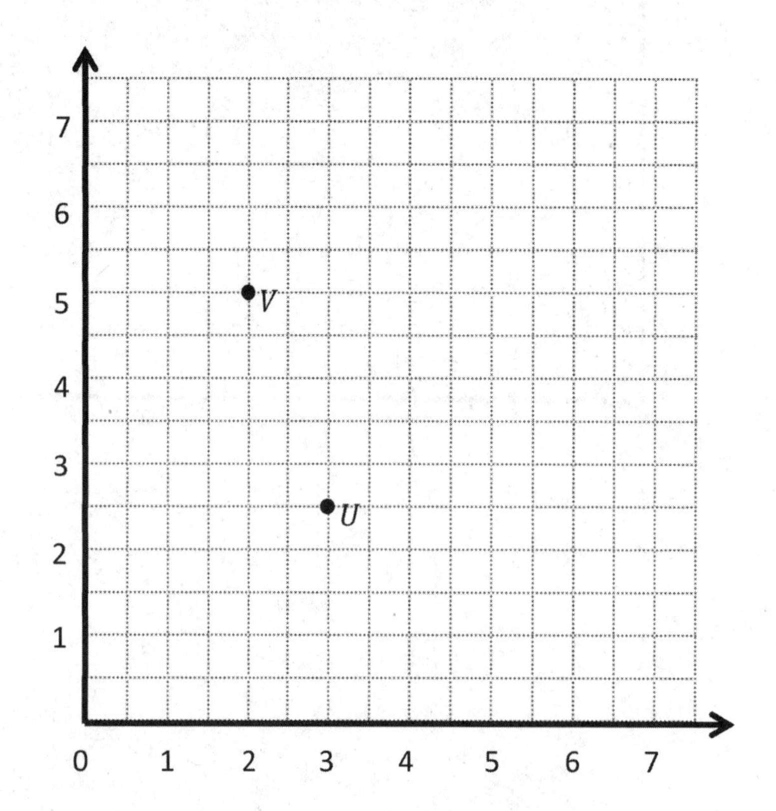

COORDINATE PLANE (CONCEPT EXPLORATION TEMPLATE)

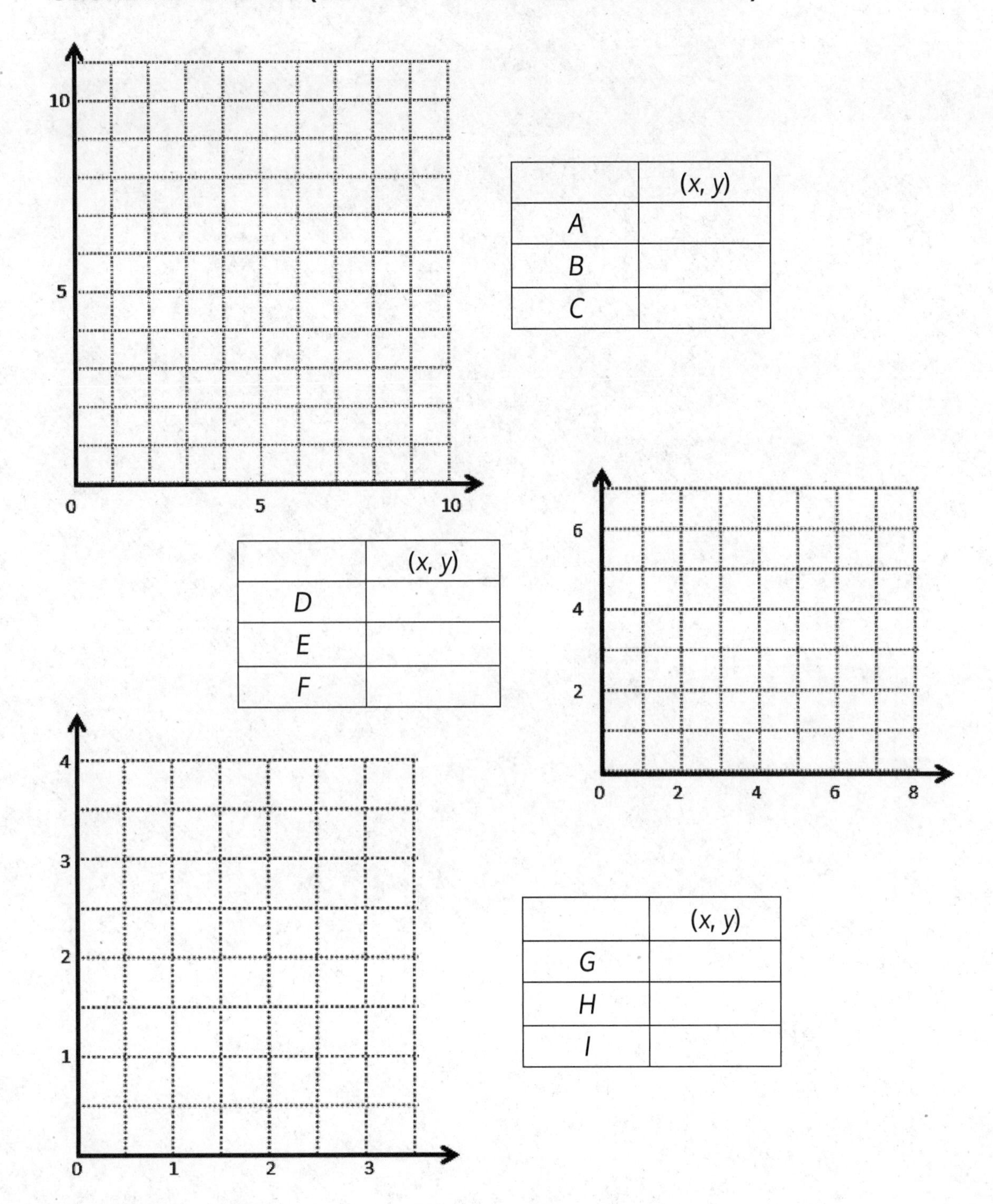

	(x, y)
A	
B	
C	

	(x, y)
D	
E	
F	

	(x, y)
G	
H	
I	

Lesson 17

Word Problem

Plot (10, 8) and (3, 3) on the coordinate plane, connect the points with a straightedge, and label them as *C* and *D*.

a. Draw a segment parallel to $\overline{CD}$.

b. Draw a segment perpendicular to $\overline{CD}$.

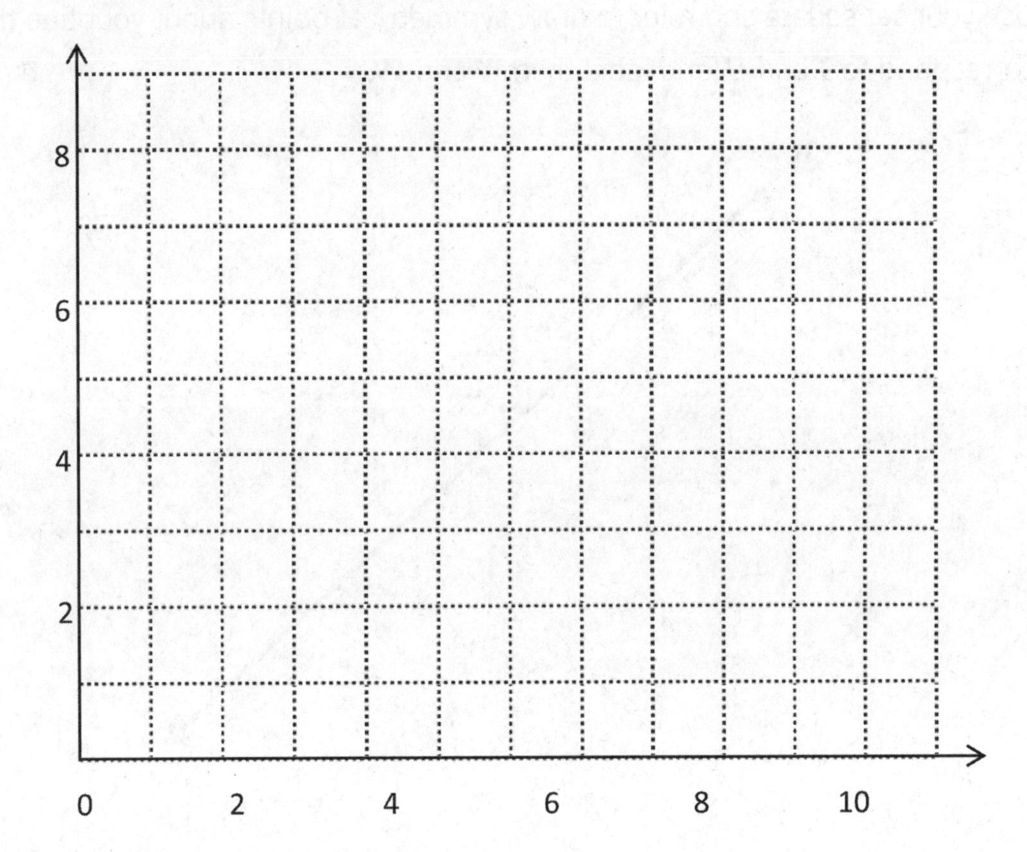

Name: ______________________________ **Date:** ____________

GRADE 5 / MISSION 6 / LESSON 17

Exit Ticket

1. Draw 2 points on one side of the line below, and label them ***T*** and ***U***.

2. Use your set square and ruler to draw symmetrical points about your line that correspond to ***T*** and ***U***, and label them ***V*** and ***W***.

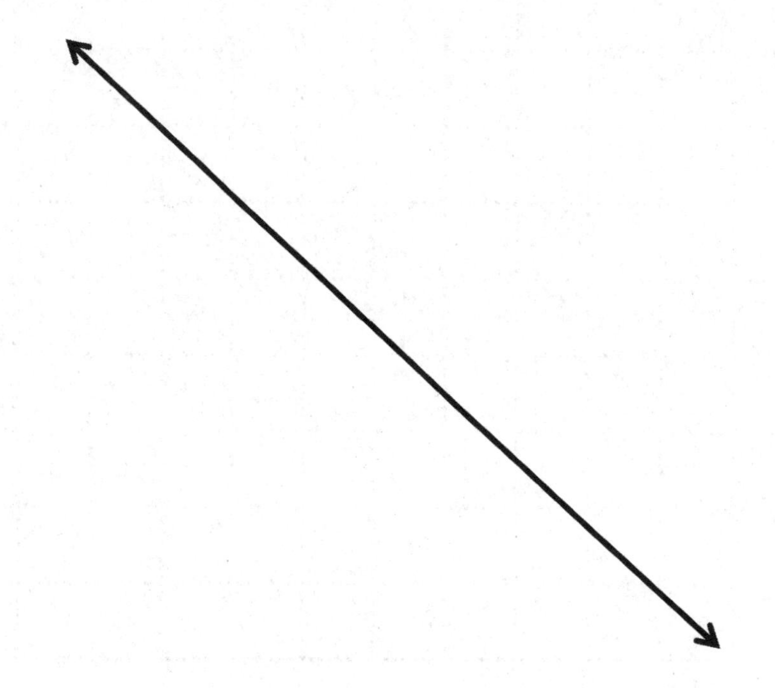

Lesson 18

Word Problem

Denis buys 8 meters of ribbon. He uses 3.25 meters for a gift. He uses the remaining ribbon equally to tie bows on 5 boxes. How much ribbon did he use on each box?

Name: ______________________ Date: ___________

GRADE 5 / MISSION 6 / LESSON 18

Exit Ticket

1. Kenny plotted the following pairs of points and said they made a symmetric figure about a line with the rule: y *is always 4.*

   (3, 2) and (3, 6)

   (4, 3) and (5, 5)

   $\left(5, \frac{3}{4}\right)$ and $\left(5, 7\frac{1}{4}\right)$

   $\left(7, 1\frac{1}{2}\right)$ and $\left(7, 6\frac{1}{2}\right)$

   Is his figure symmetrical about the line? How do you know?

COORDINATE PLANE (CONCEPT EXPLORATION TEMPLATE)

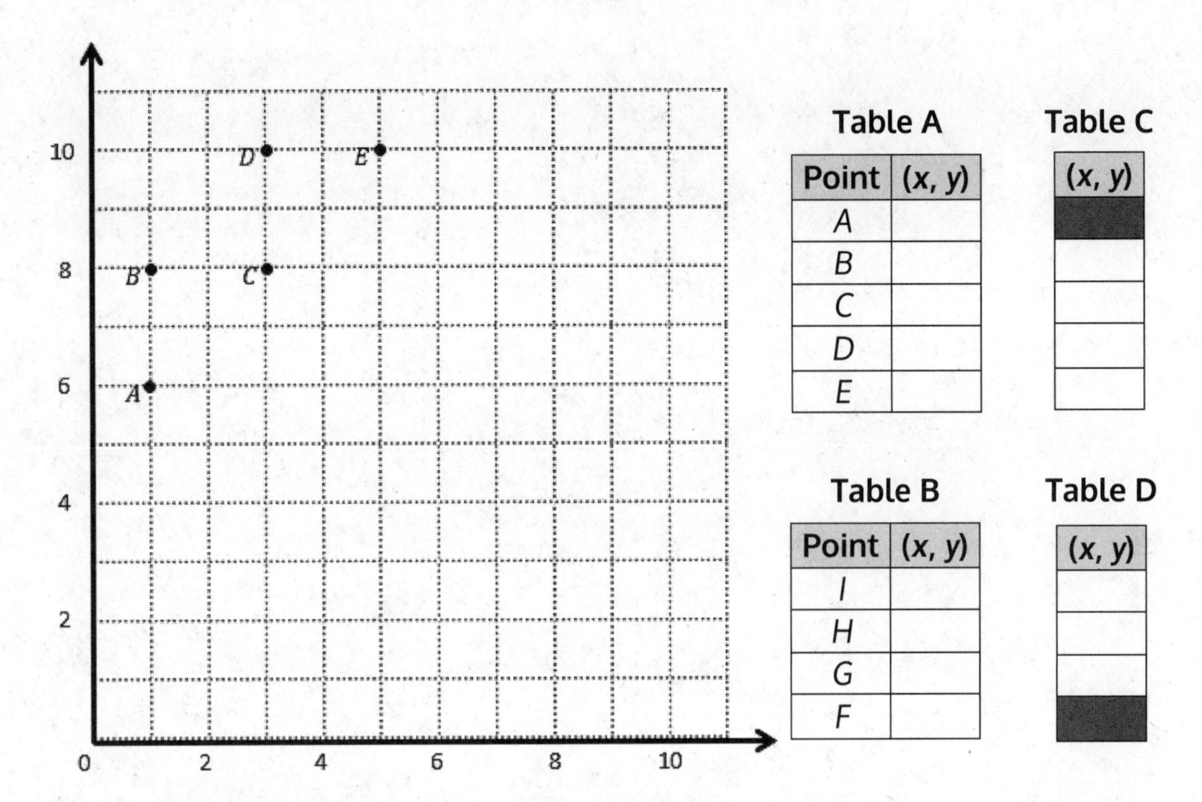

Table A

Point	(x, y)
A	
B	
C	
D	
E	

Table C

(x, y)

Table B

Point	(x, y)
I	
H	
G	
F	

Table D

(x, y)

Table E

Point	(x, y)
A	(1, 1)
B	$\left(1\frac{1}{2}, 3\frac{1}{2}\right)$
C	(2, 3)
D	$\left(2\frac{1}{2}, 3\frac{1}{2}\right)$
E	$\left(2\frac{1}{2}, 2\frac{1}{2}\right)$
F	$\left(3\frac{1}{2}, 2\frac{1}{2}\right)$
G	(3, 2)
H	$\left(3\frac{1}{2}, 1\frac{1}{2}\right)$

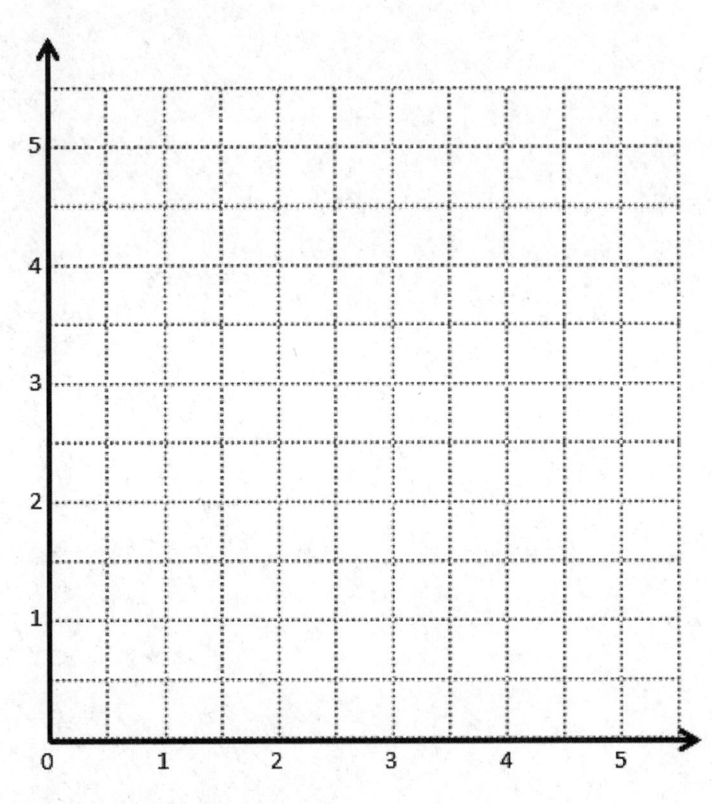

Lesson 19

Word Problem

Three feet are equal to 1 yard. The following table shows the conversion. Use the information to complete the following tasks:

Feet	Yards
3	1
6	2
9	3
12	4

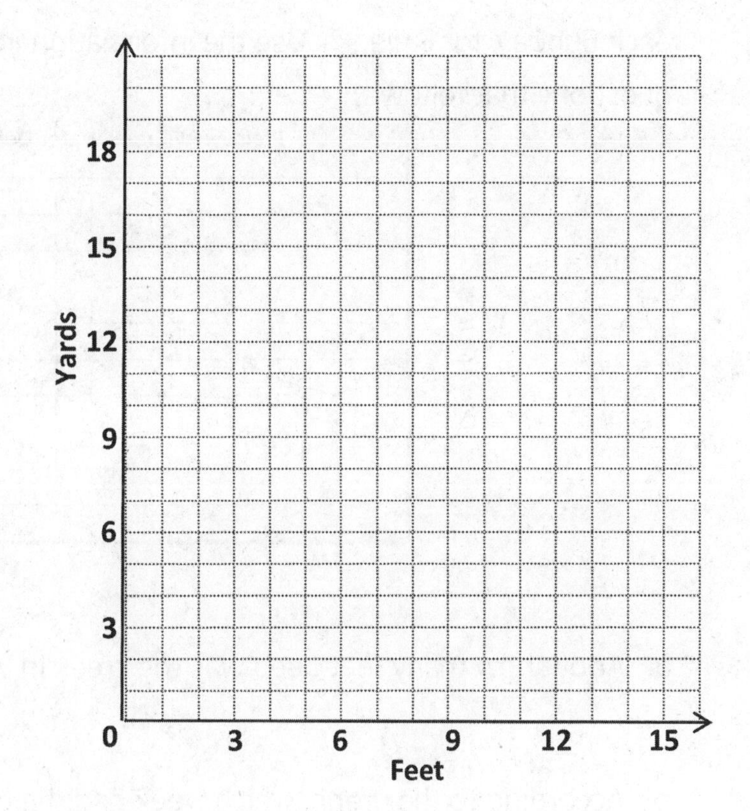

1. Plot each set of coordinates.

2. Use a straightedge to connect each point.

3. Plot one more point on this line, and write its coordinates.

4. 27 feet can be converted to how many yards? ________

5. Write the rule that describes the line.

Name: ________________________________ **Date:** ____________

GRADE 5 / MISSION 6 / LESSON 19

Exit Ticket

1. The line graph below tracks the water level of Plainsview Creek, measured each Sunday, for 8 weeks. Use the information in the graph to answer the questions that follow.

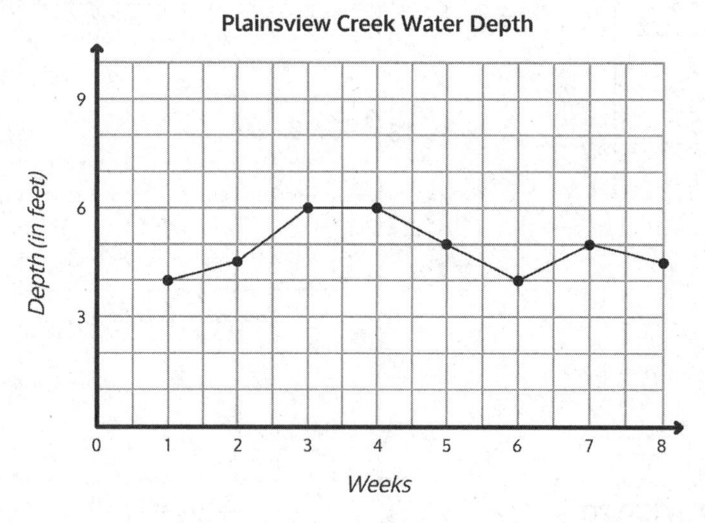

   a. About how many feet deep was the creek in Week 1? ________

   b. According to the graph, which week had the greatest change in water depth? ________

   c. It rained hard throughout the sixth week. During what other weeks might it have rained? Explain why you think so.

   d. What might have been another cause leading to an increase in the depth of the creek?

LINE GRAPH PRACTICE SHEET (CONCEPT EXPLORATION TEMPLATE)

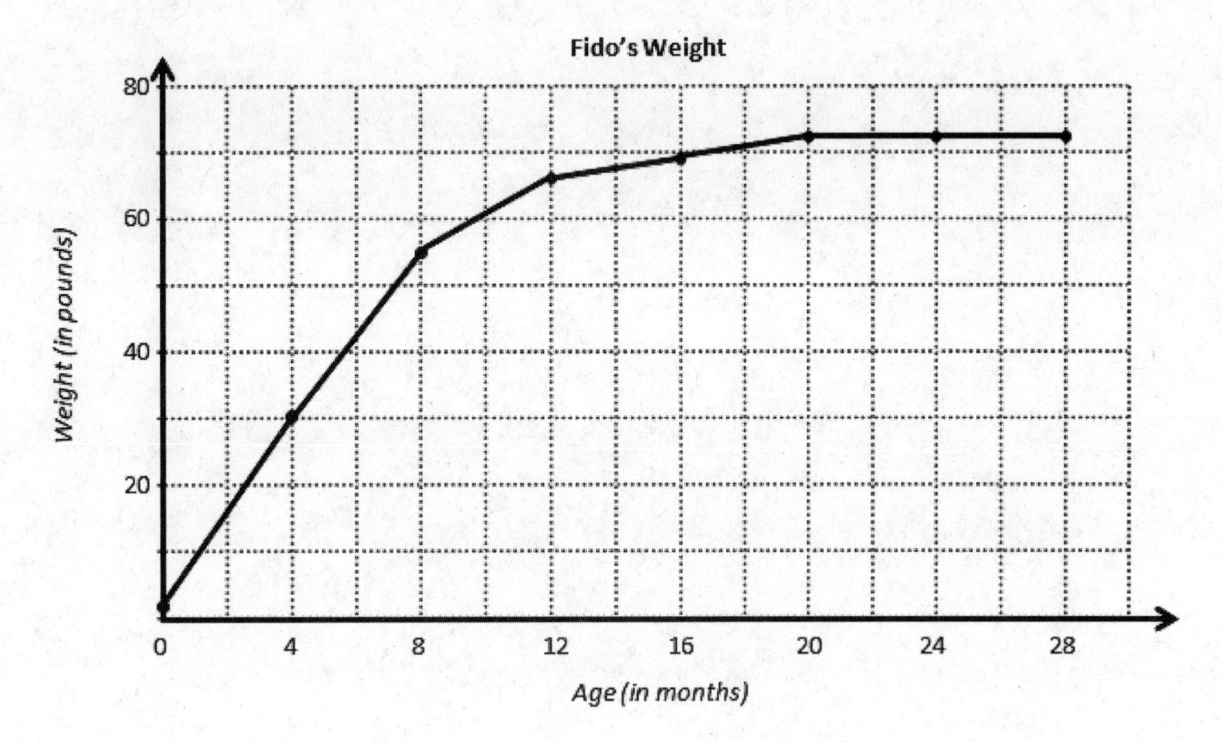

Lesson 20

Name: ______________________ Date: __________

GRADE 5 / MISSION 6 / LESSON 20

Exit Ticket

1. Read the following information about the line graph below. Then, answer the questions.

   Harry runs a hot dog stand at the county fair. When he arrived on Wednesday, he had 38 dozen hot dogs for his stand. The graph shows the number of hot dogs (in dozens) that remained unsold at the end of each day of sales.

   a. How many dozen hot dogs did Harry sell on Wednesday? How do you know?

b. Between which two-day period did the number of hot dogs sold change the most? Explain how you determined your answer.

c. During which three days did Harry sell the most hot dogs?

d. How many dozen hot dogs were sold on these three days?

PROBLEM SET

Name ______________________ Date ______________________

1. The line graph below tracks the total tomato production for one tomato plant. The total tomato production is plotted at the end of each of 8 weeks. Use the information in the graph to answer the questions that follow.

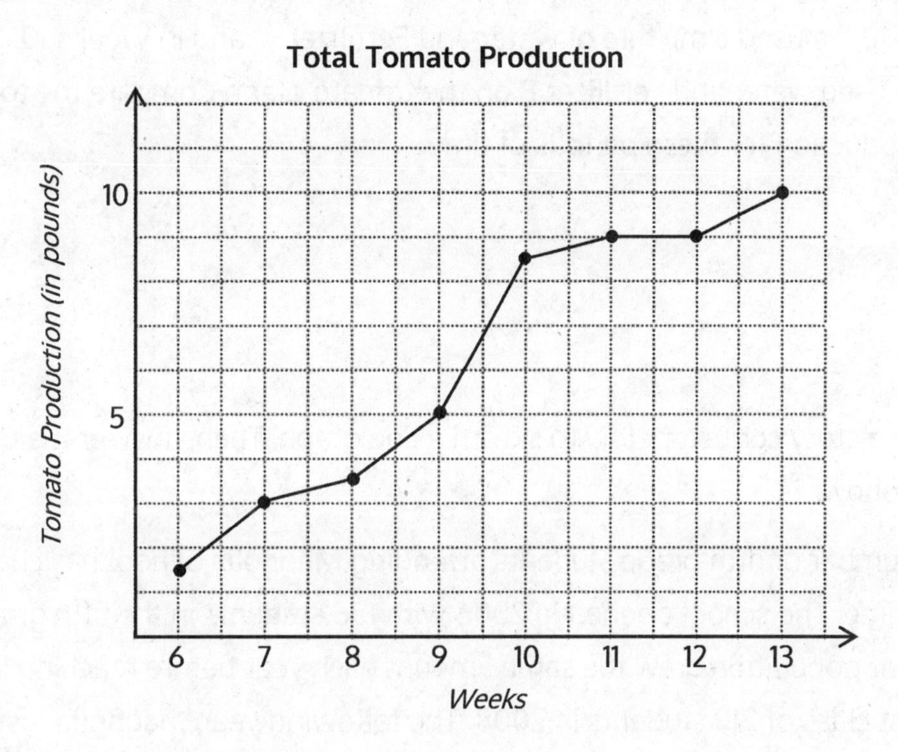

a. How many pounds of tomatoes did this plant produce at the end of 13 weeks?

b. How many pounds of tomatoes did this plant produce from Week 7 to Week 11? Explain how you know.

c. Which one-week period showed the greatest change in tomato production? The least? Explain how you know.

d. During Weeks 6–8, Jason fed the tomato plant just water. During Weeks 8–10, he used a mixture of water and Fertilizer A, and in Weeks 10–13, he used water and Fertilizer B on the tomato plant. Compare the tomato production for these periods of time.

2. Use the story context below to sketch a line graph. Then, answer the questions that follow.

The number of fifth-grade students attending Magnolia School has changed over time. The school opened in 2006 with 156 students in the fifth grade. The student population grew the same amount each year before reaching its largest class of 210 students in 2008. The following year, Magnolia lost one-seventh of its fifth graders. In 2010, the enrollment dropped to 154 students and remained constant in 2011. For the next two years, the enrollment grew by 7 students each year.

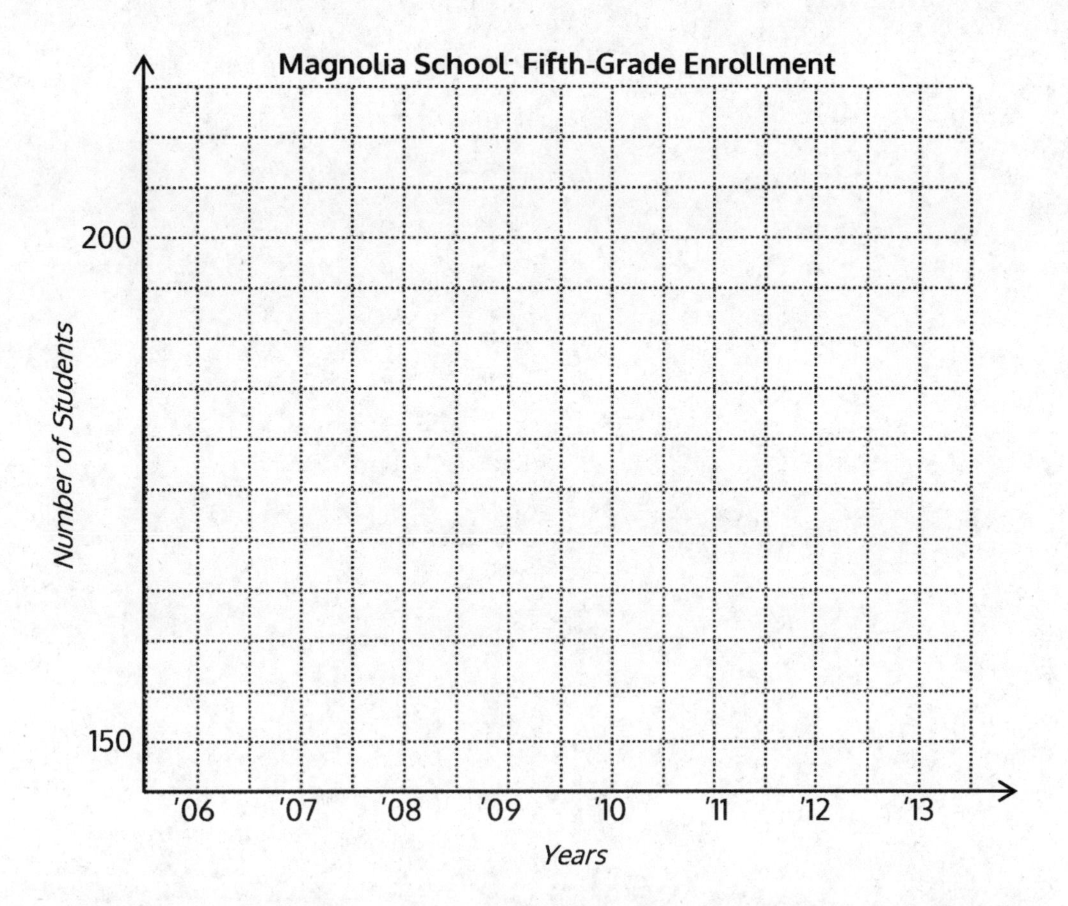

a. How many more fifth-grade students attended Magnolia in 2009 than in 2013?

b. Between which two consecutive years was there the greatest change in student population?

c. If the fifth-grade population continues to grow in the same pattern as in 2012 and 2013, in what year will the number of students match 2008's enrollment?

Lesson 26

Word Problem

The market sells watermelons for $0.39 per pound and apples for $0.43 per pound. Write an expression that shows how much Carmen spends for a watermelon that weighs 11.5 pounds and a bag of apples that weighs 3.2 pounds.

Name: ______________________________ Date: ____________

GRADE 5 / MISSION 6 / LESSON 26

Reflection

How did the games we played today prepare you to practice writing, solving, and comparing expressions? Why do you think these are important skills to work on? Will you teach someone at home how to play these games with you? What math skills will you need to teach in order for someone at home to be able to play with you?

COMPARING EXPRESSIONS GAME BOARD (CONCEPT EXPLORATION TEMPLATE 2)

$96 \times \left(63 + \frac{17}{12}\right)$	◯	$(96 \times 63) + \frac{17}{12}$
$\left(437 \times \frac{9}{15}\right) \times \frac{6}{8}$	◯	$\left(437 \times \frac{9}{15}\right) \times \frac{7}{8}$
$4 \times 8.35 + 4 \times 6.21$	◯	4×15.87
$\frac{6}{7} \times (3{,}065 + 4{,}562)$	◯	$(3{,}065 + 4{,}562\ ) + \frac{6}{7}$
$(8.96 \times 3) + (5.07 \times 8)$	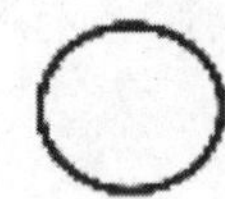	$(8.96 + 3) \times (5.07 + 8)$
$\left(297 \times \frac{16}{15}\right) + \frac{8}{3}$	◯	$\left(297 \times \frac{13}{15}\right) + \frac{8}{3}$
$\frac{12}{7} \times \left(\frac{5}{4} + \frac{5}{9}\right)$	◯	$\frac{12}{7} \times \frac{5}{4} + \frac{12}{7} \times \frac{5}{9}$

Lesson 27

Name: ______________________ **Date:** ____________

GRADE 5 / MISSION 6 / LESSON 27

Reflection

How did teaching other students how to solve a word problem strengthen your skills as a problem solver?

What did you learn about your problem-solving skills?

What are your strengths and weaknesses as a problem solver?

Lesson 28

Name: ____________________ **Date:** __________

GRADE 5 / MISSION 6 / LESSON 28

Reflection

What math skills have you improved through our Fluency Practice?

How do you know you've improved?

What math skills do you need to continue to practice? Why?

PROBLEM SET

Name ____________________ Date ____________________

1. Answer the following questions about fluency.

   a. What does being fluent with a math skill mean to you?

   b. Why is fluency with certain math skills important?

   c. With which math skills do you think you should be fluent?

   d. With which math skills do you feel most fluent? Least fluent?

   e. How can you continue to improve your fluency?

2. Use the chart below to list skills from today's activities with which you are fluent.

Fluent Skills

3. Use the chart below to list skills we practiced today with which you are less fluent.

Skills to Practice More

Lesson 29

Name: ______________________________ **Date:** ____________

GRADE 5 / MISSION 6 / LESSON 29

Reflection

It is said that the true measure of knowing something is being able to teach it to someone else. Who can you teach these terms to? How will you teach these terms to your student?

Lesson 30

Name: ______________________ **Date:** ____________

GRADE 5 / MISSION 6 / LESSON 30

Reflection

Playing math games can be a fun way to practice math skills. How will you use the games to retain these terms? Who will play with you? How can you change the games to play alone? How often will you play games?

Lesson 31

Word Problem

You will need a protractor and a ruler for this problem.

Step 1 Draw $\overline{AB}$ 3 inches long centered near the bottom of the page.

Step 2 Draw $\overline{AC}$ 3 inches long, such that $\angle BAC$ measures 108°.

Step 3 Draw $\overline{CD}$ 3 inches long, such that $\angle ACD$ measures 108°.

Step 4 Draw $\overline{DE}$ 3 inches long, such that $\angle CDE$ measures 108°.

Step 5 Draw $\overline{EB}$.

Step 6 Measure $\overline{EB}$.

What is the length of $\overline{EB}$?

What shape have you drawn?

PROBLEM SET

Name ______________________ Date ______________________

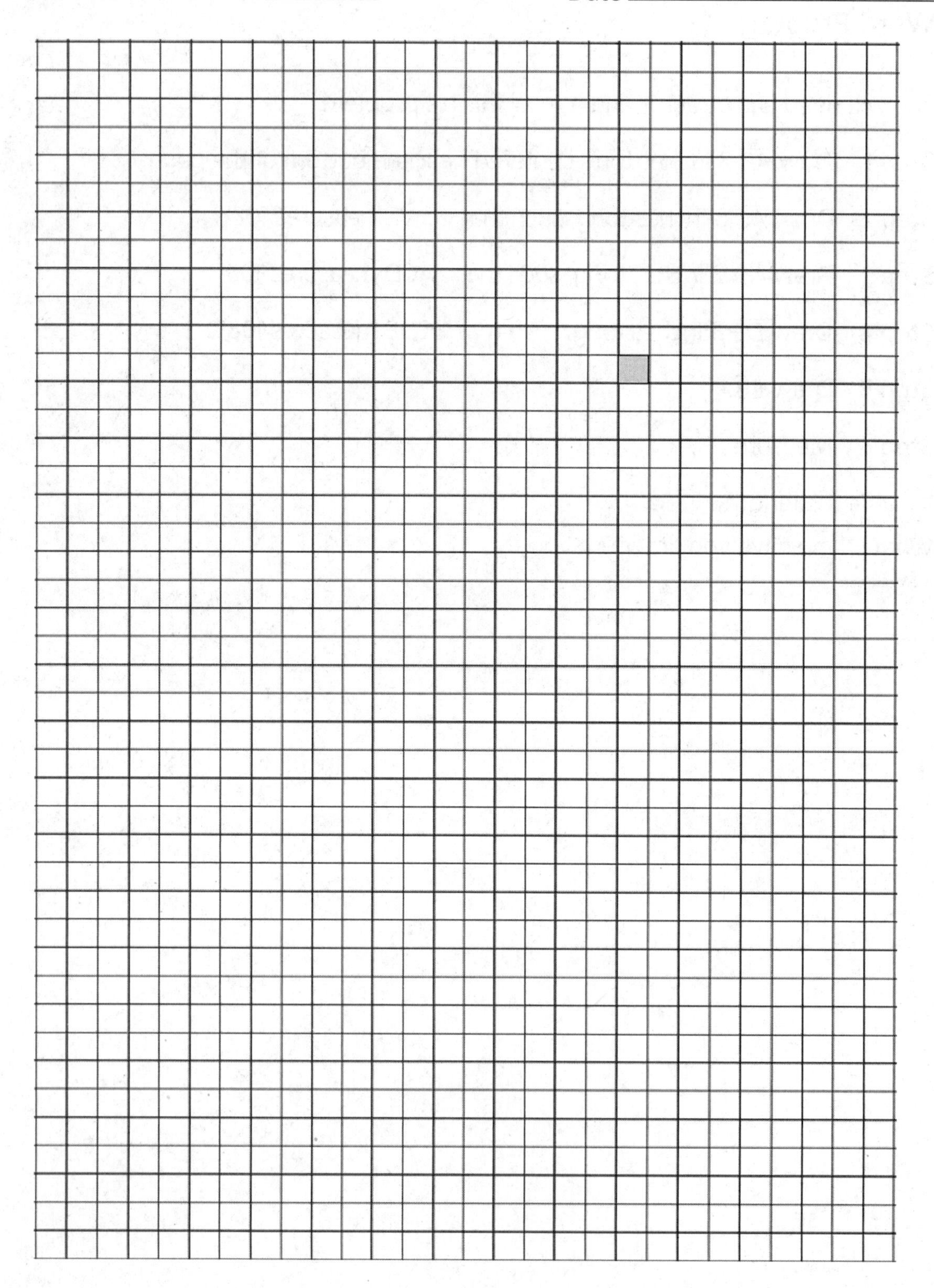

Lesson 32

Word Problem

Look at the Fibonacci sequence you just wrote. Analyze which numbers are even. Is there a pattern to the even numbers? Why? Think about the spiral of squares that you made yesterday.

Name: ____________________ Date: __________

GRADE 5 / MISSION 6 / LESSON 32

Reflection

Today, we watched how savings can grow over time, but we did not discuss how the money saved was earned. Have you ever thought about how math skills might help you to earn money? If so, what are some jobs that might require strong math skills? If not, think about it now. How might you make a living using math skills?

PROBLEM SET

Name ______________________ Date ______________________

1. Ashley decides to save money, but she wants to build it up over a year. She starts with $1.00 and adds 1 more dollar each week. Complete the table to show how much she will have saved after a year.

Week	Add	Total
1	$1.00	$1.00
2	$2.00	$3.00
3	$3.00	$6.00
4	$4.00	$10.00
5		
6		
7		
8		
9		
10		
11		
12		
13		
14		
15		
16		
17		
18		
19		
20		
21		
22		
23		
24		
25		
26		

Week	Add	Total
27		
28		
29		
30		
31		
32		
33		
34		
35		
36		
37		
38		
39		
40		
41		
42		
43		
44		
45		
46		
47		
48		
49		
50		
51		
52		

2. Carly wants to save money, too, but she has to start with the smaller denomination of quarters. Complete the second chart to show how much she will have saved by the end of the year if she adds a quarter more each week. Try it yourself, if you can and want to!

Week	Add	Total
1	\$0.25	\$0.25
2	\$0.50	\$0.75
3	\$0.75	\$1.50
4	\$1.00	\$2.50
5		
6		
7		
8		
9		
10		
11		
12		
13		
14		
15		
16		
17		
18		
19		
20		
21		
22		
23		
24		
25		
26		

Week	Add	Total
27		
28		
29		
30		
31		
32		
33		
34		
35		
36		
37		
38		
39		
40		
41		
42		
43		
44		
45		
46		
47		
48		
49		
50		
51		
52		

3. David decides he wants to save even more money than Ashley did. He does so by adding the next Fibonacci number instead of adding $1.00 each week. Use your calculator to fill in the chart and find out how much money he will have saved by the end of the year. Is this realistic for most people? Explain your answer.

Week	Add	Total
1	\$1	\$1
2	\$1	\$2
3	\$2	\$4
4	\$3	\$7
5	\$5	\$12
6	\$8	\$20
7		
8		
9		
10		
11		
12		
13		
14		
15		
16		
17		
18		
19		
20		
21		
22		
23		
24		
25		
26		

Week	Add	Total
27		
28		
29		
30		
31		
32		
33		
34		
35		
36		
37		
38		
39		
40		
41		
42		
43		
44		
45		
46		
47		
48		
49		
50		
51		
52		

Lesson 33

PROBLEM SET

Name ______________________ Date ______________________

Record the dimensions of your boxes and lid below. Explain your reasoning for the dimensions you chose for Box 2 and the lid.

BOX 1 (Can hold Box 2 inside.) The dimensions of Box 1 are ________ × ________ × ________. Its volume is ________.
BOX 2 (Fits inside of Box 1.) The dimensions of Box 2 are ________ × ________ × ________. Reasoning:
LID (Fits snugly over Box 1 to protect the contents.) The dimensions of the lid are ________ × ________ × ________. Reasoning:

1. What steps did you take to determine the dimensions of the lid?

2. Find the volume of Box 2. Then, find the difference in the volumes of Boxes 1 and 2.

3. Imagine Box 3 is created such that each dimension is 1 cm less than that of Box 2. What would the volume of Box 3 be?

Lesson 34

Word Problem

Steven is a ____________________ who had \$280. He spent $\frac{1}{4}$ of his money on a ____________________ and $\frac{5}{6}$ of the remainder on a ____________________. How much money did he spend altogether?

PROBLEM SET

Name ______________________ Date ______________________

I reviewed ________________'s work.

Use the chart below to evaluate your friend's two boxes and lid. Measure and record the dimensions, and calculate the box volumes. Then, assess suitability, and suggest improvements in the adjacent columns.

Dimensions and Volume	Is the Box or Lid Suitable? Explain.	Suggestions for Improvement
BOX 1 dimensions: Total volume:		
BOX 2 dimensions: Total volume:		
LID dimensions: